Mosby's Systems Atlas of Human Anatomy

Mosby–Year Book, Inc.
11830 Westline Industrial Drive
St. Louis, MO 63146

EAI

Engineering Animation, Inc.
2321 North Loop Drive
Ames, IA 50010

Publisher: **James Smith**

Editor: **Ron Worthington, Ph. D.**

Manufacturing Manager: **Theresa Fuchs**

Project Manager: **Dana Peick**

Cover & Book Production: **Engineering Animation, Inc**

Printed in the United States of America

Mosby–Year Book, Inc.
11830 Westline Industrial Drive
St. Louis, MO 63146

ISBN 0-8151-8654-1

Foreword

In its 1985 Long Range Plan, the Board of Regents of the National Library of Medicine challenged the NLM staff to investigate, thoroughly and systematically, the technical requirements for and feasibility of instituting a biomedical images library. The staff of the Lister Hill National Center for Biomedical Communications, NLM's research and development division, took up the challenge. To explore the possibilities, many questions were raised: What are the differences between electronic digital images versus printed images? In the development of a biomedical image library should the images be stored in an electronic digital format? What can be done with an electronic digital image that cannot be done with a printed image? Which areas of biomedicine would best demonstrate the potential benefits of electronic digital images?

Anatomy, a foundation of biomedicine, was determined to be the area with the most potential for this effort because it is clearly visual and largely image based. At the 1988 NLM-sponsored Workshop on Digital Imaging it was noted that although anatomy was three-dimensional and a knowledge of anatomy required spatial understanding, anatomy was learned, except in the cadaver laboratory, through flat, two-dimensional images. Electronic digital anatomical images had the potential to change this. The NLM's Board of Regents agreed and accepted the recommendations in its 1990 Report of the Board of Regents on Electronic Imaging that NLM should undertake for the first time a project building a digital-image library of volumetric data representing a complete, normal adult male and a complete, normal adult female. This *Visible Human Project* will include digitized photographic images for cryosectioning, digital images derived from computerized tomography, and digital magnetic resonance images of cadavers. Five years later this vision has become reality.

Mosby's Systems Atlas of Human Anatomy demonstrates many of the reasons the visionaries of the late 1980s argued that anatomical atlases would be enhanced through the use of electronic digital images and digital-imaging technology. An illustrated image is flat and students must imagine the three-dimensional relationships trapped in the illustration. The images of Mosby's Systems Atlas of Human Anatomy are given a three-dimensional appearance through volume rendering of the digital-image data. A volume-rendered image provides a sense of depth and realism through the use of optical properties such as shading, shadowing, and transparency.

Anatomists understand that to study a structure it has to be isolated from its neighboring tissue. But in doing so, the relationship between the structure and its neighbors is lost. Often the *in situ* shape of the organ is also distorted. With a volume-rendered anatomical atlas, each organ can be isolated, yet its *in situ* orientation and shape are preserved. The organ can be viewed from every conceivable orientation because neighboring tissue can be made to disappear or can become translucent so that the student can visualize the anatomical relationships without distortion.

Mosby's Systems Atlas of Human Anatomy is the first of a new generation of multimedia anatomical atlases. It clearly demonstrates the potential of digital imagery and volume-rendering for studying anatomy.

Michael J. Ackerman, Ph.D.
Project Officer,
Visible Human Project

This article was written by the author in his private capacity. No official support or endorsement by the National Library of Medicine is intended or should be inferred.

Acknowledgments

Engineering Animation, Inc. (EAI) and Mosby–Year Book would like to acknowledge the National Library of Medicine (NLM) for the Visible Human Project datasets. We commend the NLM for achieving the vision of a digital biomedical image library. The color cross-sectional images have become the standard dataset for creating quality three-dimensional renderings of the human anatomy. The volume-renderings found in this book have been produced exclusively from the Visible Human male dataset. Visit the NLM homepage at *http://www.nlm.nih.gov* for more information on the National Library of Medicine and the Visible Human Project.

EAI and Mosby would like to further acknowledge Hewlett-Packard for providing the workstations used to implement EAI's volume visualization algorithms and to render the images found in this atlas.

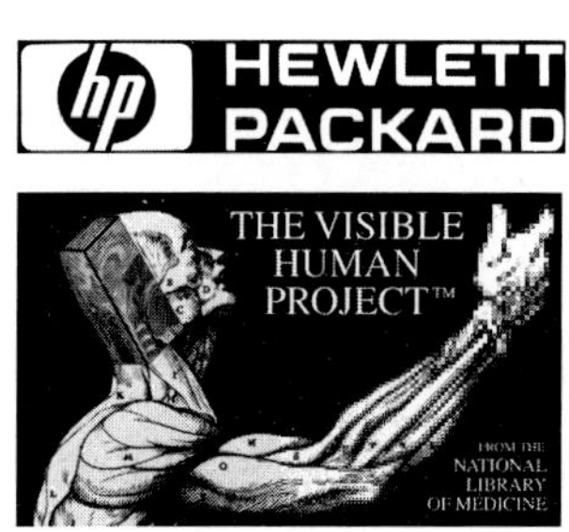

Mosby's Systems Atlas of Human Anatomy

Table of Contents

Skeletal System **1**

Skeletal System 1 2
Skeletal System 2 3
Skull 4
Base of Skull 1 5
Base of Skull 2 6
Mandible 7
Spine 8
Spine with Pelvis 9
Atlas and Axis 10
Cervical Vertebrae 11
Thoracic Vertebrae 12
Lumbar Vertebrae 13
Sacrum and Coccyx 14
Ribcage 15
Sternum 16
Rib 17
Upper Appendicular Skeleton 1 18
Upper Appendicular Skeleton 2 19
Pectoral Girdle 1 20
Pectoral Girdle 2 21
Upper Extremity 1: Humerus and Forearm 22
Upper Extremity 2: Radius and Ulna 23
Upper Extremity 3: Hand 24
Lower Appendicular Skeleton 25
Pelvis 26
Innominate Bone, Sacrum and Coccyx 27
Lower Extremity 1: Femur 28
Lower Extremity 2: Femur and Patella 29
Lower Extremity 3: Tibia and Fibula 30
Lower Extremity 4: Tibia and Fibula 31
Lower Extremity 5: Foot 32

Muscular System **33**

Muscular System 1 34
Muscular System 2 35
Muscles of the Head and Neck 1 36
Muscles of the Head and Neck 2 37
Torso 1 38
Torso 2 39
Muscles of the Back 40
Abdominal Muscles 1 41
Abdominal Muscles 2 42
Muscles of the Upper Extremity 1 43
Muscles of the Upper Extremity 2 44
Muscles of the Upper Extremity 3 45
Cross-sections of the Shoulder 46
Muscles of the Shoulder and Chest 47
Deep Muscles of the Upper Extremity 1 48
Deep Muscles of the Upper Extremity 2 49
Muscles of the Forearm 50
Cross-sections of the Hip 51
Muscles of the Lower Extremity 1 52
Muscles of the Lower Extremity 2 53
Muscles of the Thigh 54
Muscles of the Leg 55
Deep Muscles of the Lower Extremity 56
Cross-sections of the Knee 57
Cross-sections of the Ankle 58

Nervous System **59**

Nervous System 1 60
Nervous System 2 61
Brain 1 62
Brain 2 63
Brain 3 64
Ventricles of the Brain 65
Cerebellum 66
Axial Sections of the Brain 67
Sagittal Sections of the Brain 68
Coronal Sections of the Brain 69
Brachial Plexus 70
Lumbosacral Plexus 71

Cardiovascular System **72**

Cardiovascular System 1:Systemic Circulation 73
Cardiovascular System 2:Systemic Circulation 74
Heart 75
Sections Through the Heart 1 76
Sections Through the Heart 2 77
Cross-sections of the Heart 1 78
Cross-sections of the Heart 2 79
Vessels of the Head and Neck 1 80
Vessels of the Head and Neck 2 81
Great Vessels 82
Abdominal Vasculature 83
Bronchial Arteries and Veins 1 84
Bronchial Arteries and Veins 2 85
Hepatic Portal System 1 86
Hepatic Portal System 2 87

Lymphatic System **88**

Lymphatic System 89
Spleen and Thymus 90

Respiratory System **91**

Respiratory System 1 92
Respiratory System 2 93
Lungs 1 94
Lungs 2 95
Airways 96

Alimentary System **97**

Alimentary System 1 98
Alimentary System 2 99
Upper Gastrointestinal Tract 100
Small Intestine 101
Large Intestine 102
Pancreas 103
Liver 1 104
Liver 2 105
Cross-sections of the Abdomen 1 106
Cross-sections of the Abdomen 2 107

Urinary System ***108***
Urinary System 1 109
Urinary System 2 110
Cross-sections of the Kidney 111

Reproductive System ***112***
Male Reproductive System 113
Male Genital Organs 114
Cross-sections of the Male Pelvic Organs 115

Endocrine System ***116***
Endocrine System 117
Adrenal Glands 118

Index 119

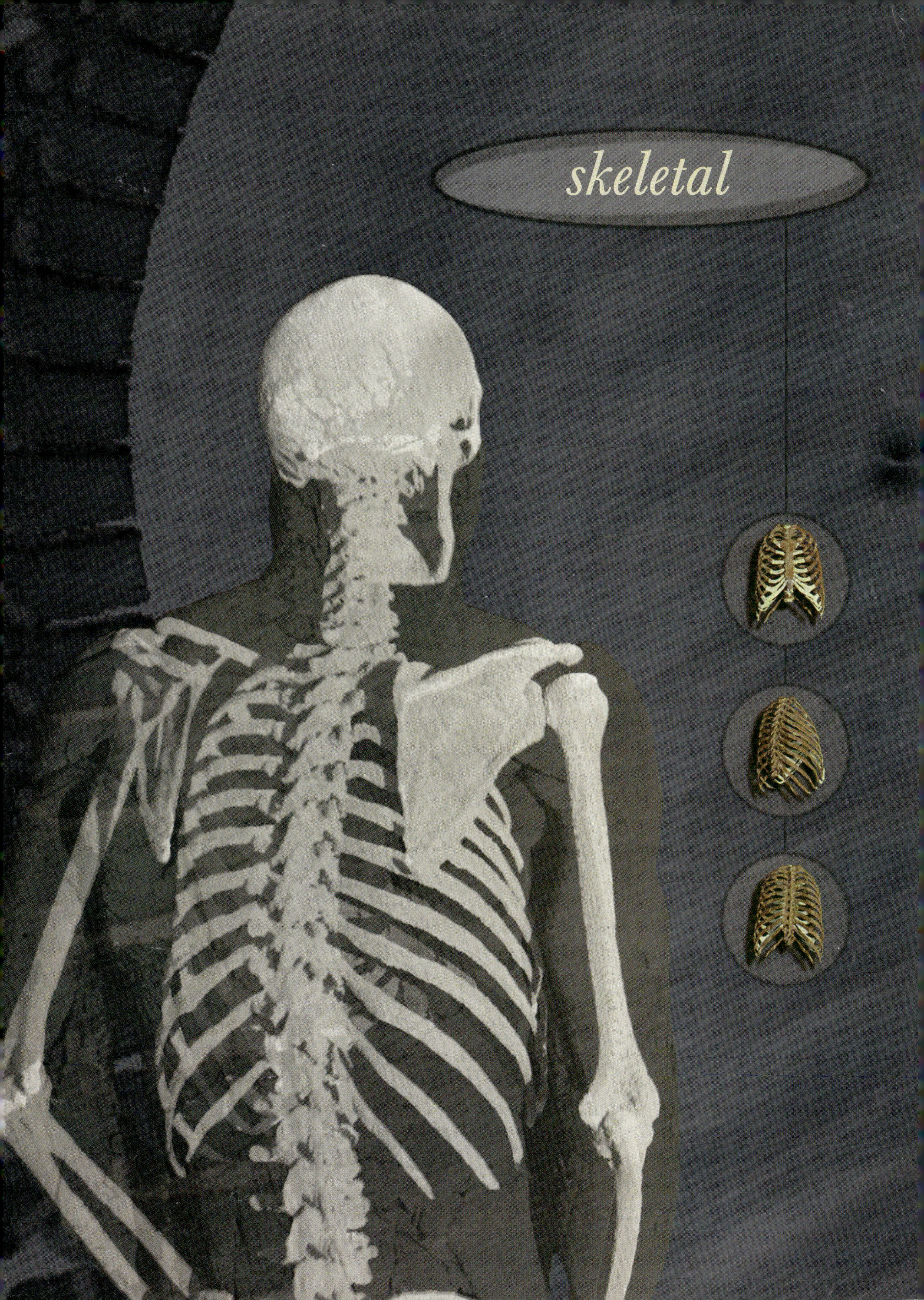
skeletal

Skeletal System 1

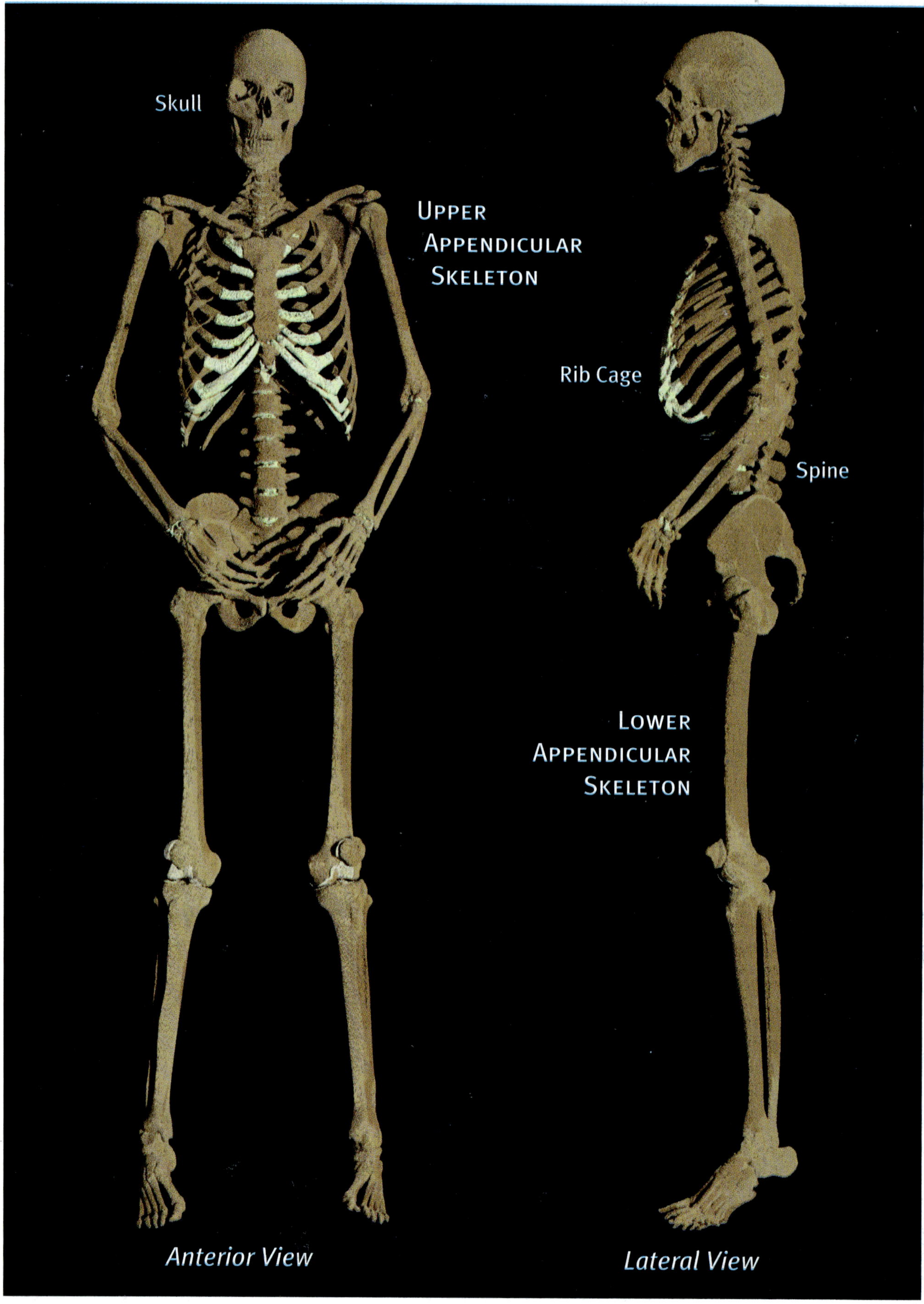

Skeletal System 2

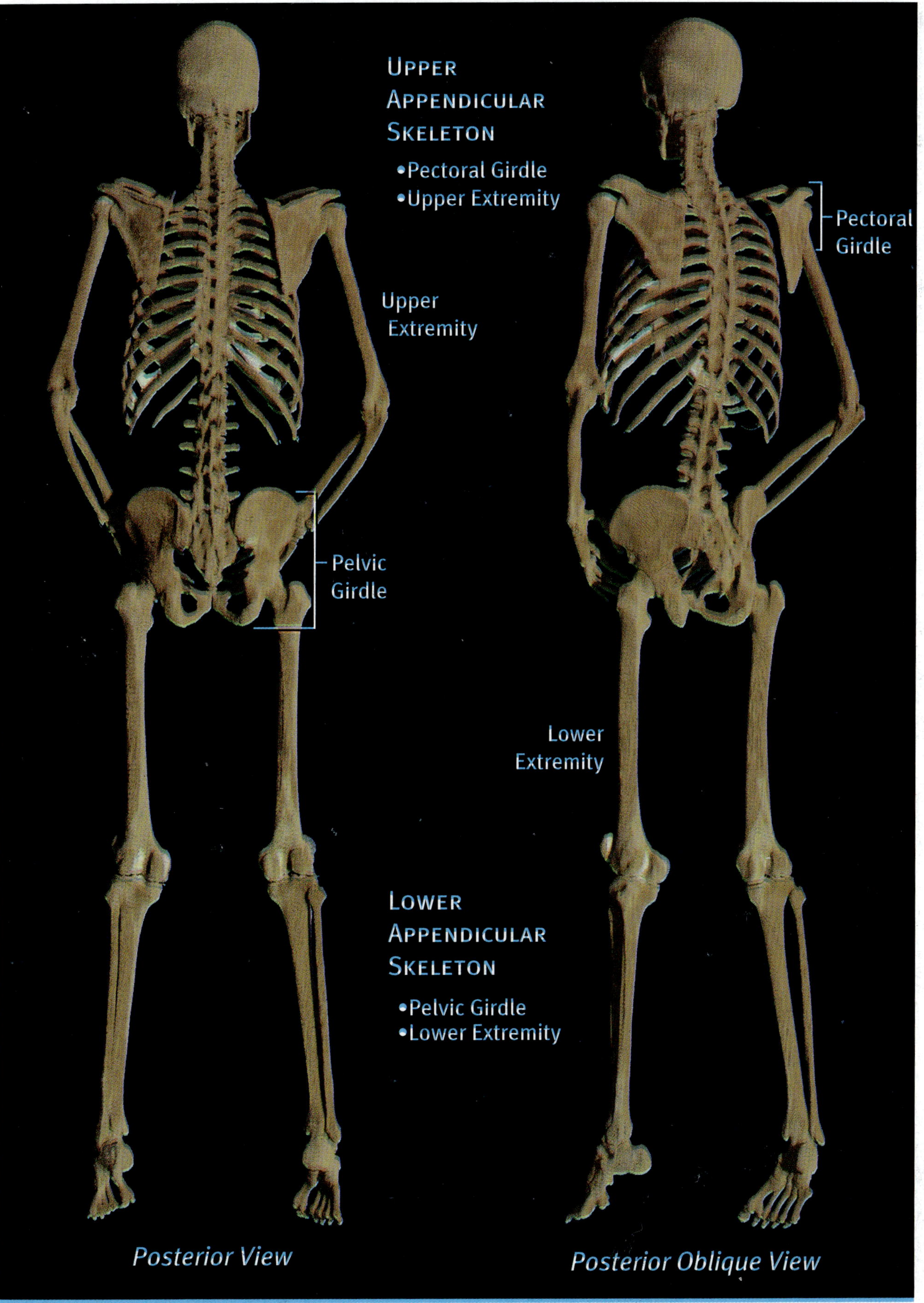

Posterior View

Posterior Oblique View

Skull

frontal bone
supraorbital notch (foramen)
orbital surface of frontal bone
lesser wing of sphenoid
superior orbital fissure
inferior orbital fissure
greater wing of sphenoid
zygomatic bone
orbital surface of maxilla
vomer
Anterior View
parietal bone
glabella
zygomatic bone
nasal bone
external acoustic meatus
temporal process
infraorbital foramen
external occipital protuberance
anterior nasal spine
styloid process of temporal bone
Lateral View
mandible
mental foramen

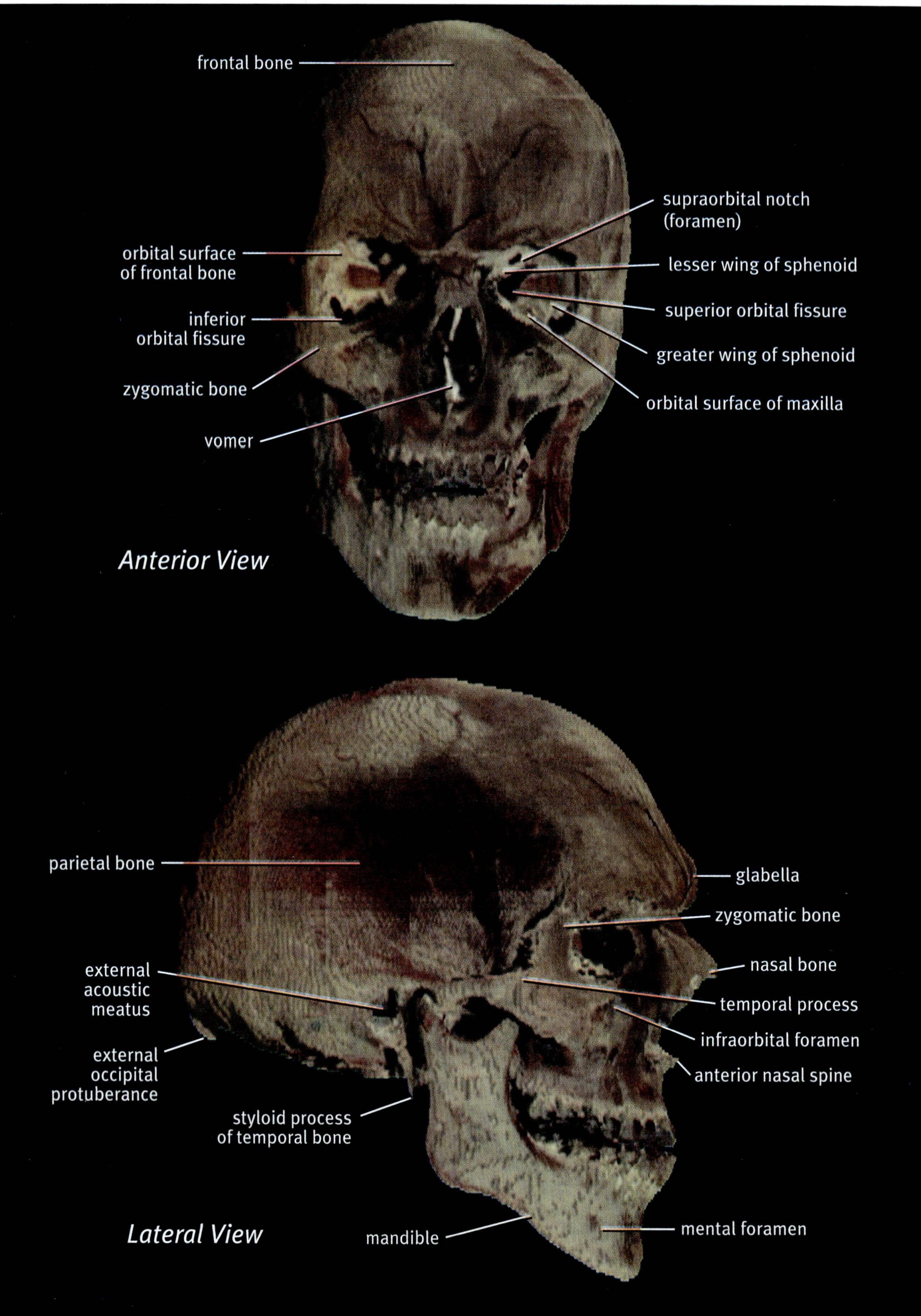

Base of Skull 1

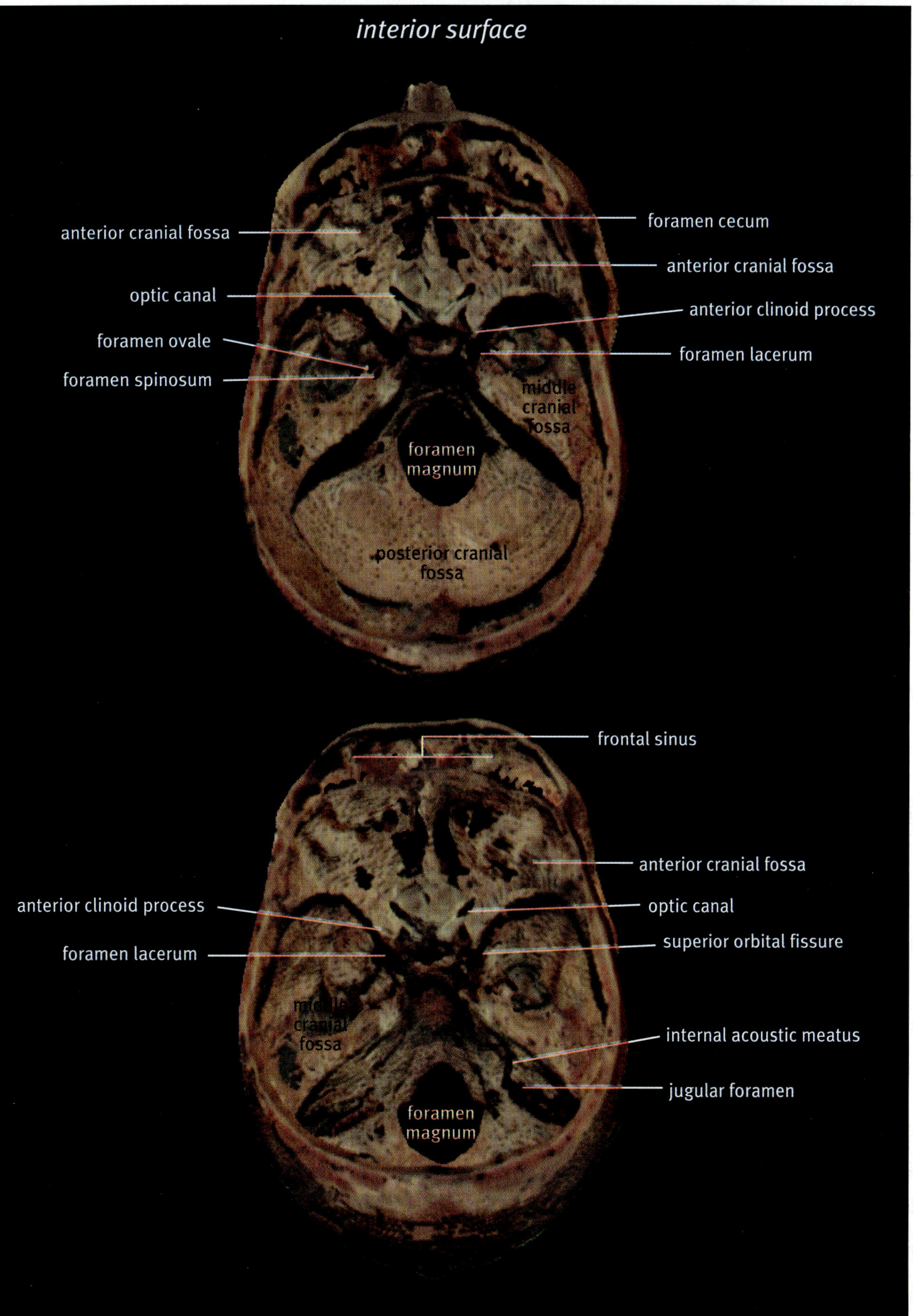

Base of Skull 2

exterior surface

maxilla
zygomatic bone
zygomatic process
zygomatic arch
hamulus of sphenoid
vomer
pharyngeal tubercle
occipital condyle
condylar canal & fossa
external occipital crest
incisive foramen
greater palatine foramina
posterior nasal spine
choanae
foramen lacerum
mandibular fossa
styloid process
mastoid process
stylomastoid process
foramen magnum
external occipital protuberance

Inferior Posterior Oblique View

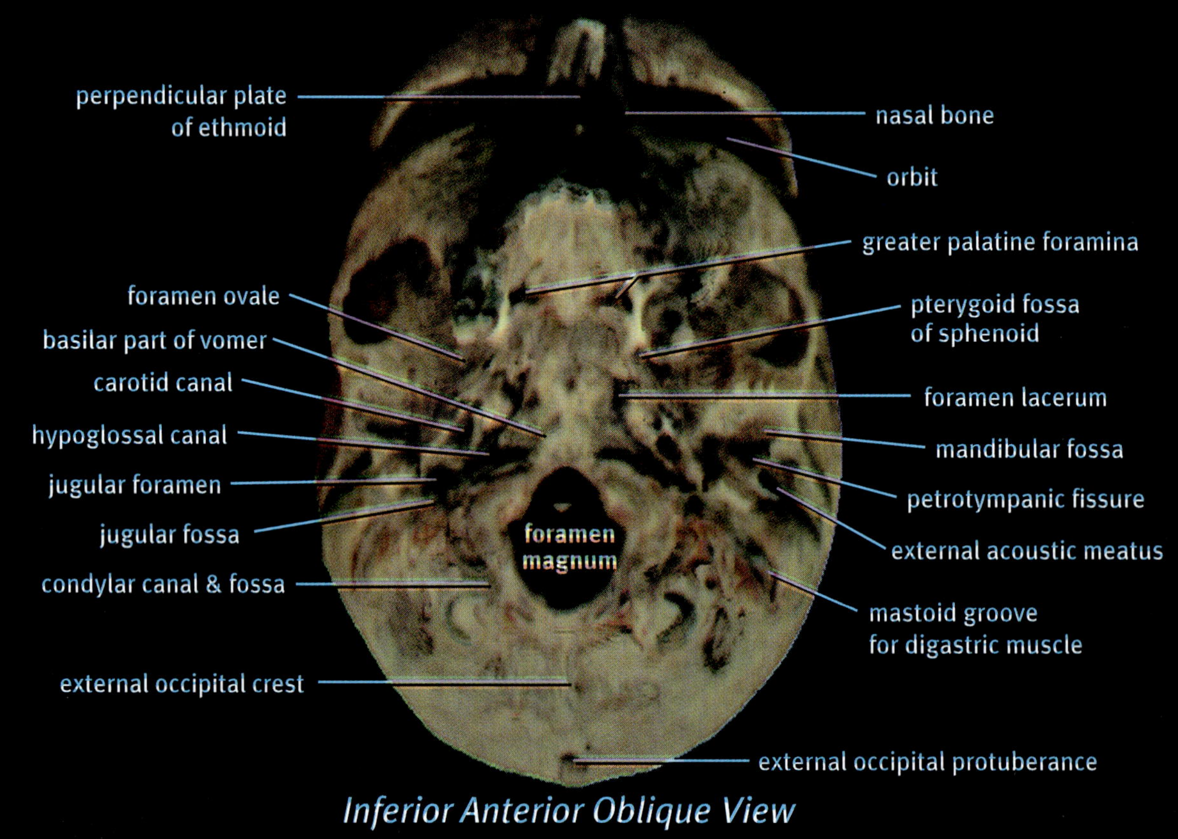

Inferior Anterior Oblique View

Mandible

Lateral View
ramus
oblique line
angle
body
Superior View
coronoid process
Anterior Oblique View
coronoid process
lateral pterygoid muscle attachment
angle
mental foramen
mental protuberance
mental tubercle
Anterior View
ramus
angle
body
mental tubercle
Inferior View
head of coronoid process
neck of coronoid process
notch of mandible
Posterior View
medial pterygoid insertion
mylohyoid line
genioglossus and geniohyoid attachment
mylohyoid attachment

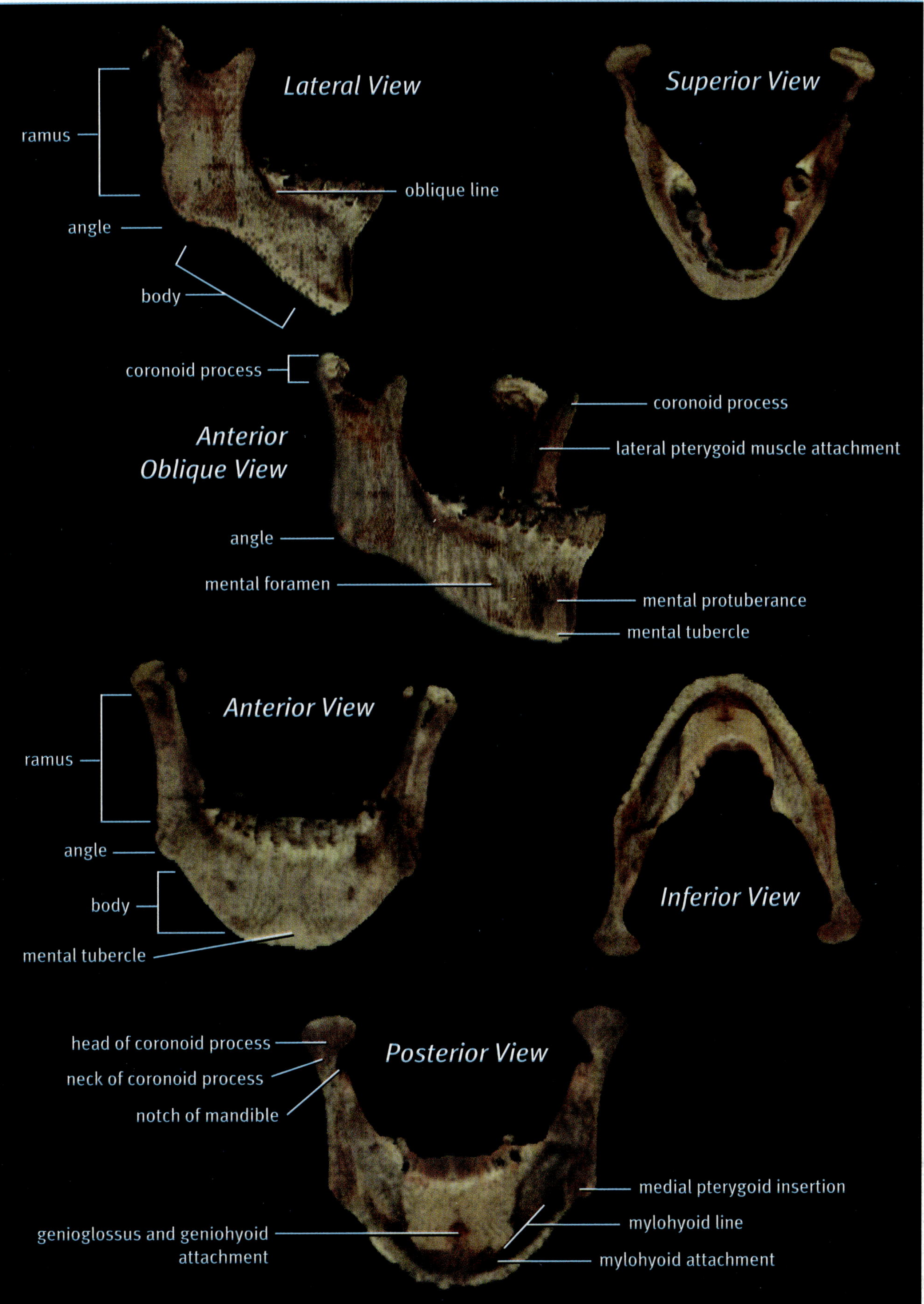

Spine

C1 (atlas)
C2 (axis)
Cervical
Vertebrae
C7
T1
Thoracic
Vertebrae
T12
L1
Lumbar
Vertebrae
L5
sacrum
coccyx
Anterior Oblique View
Lateral View
Posterior View

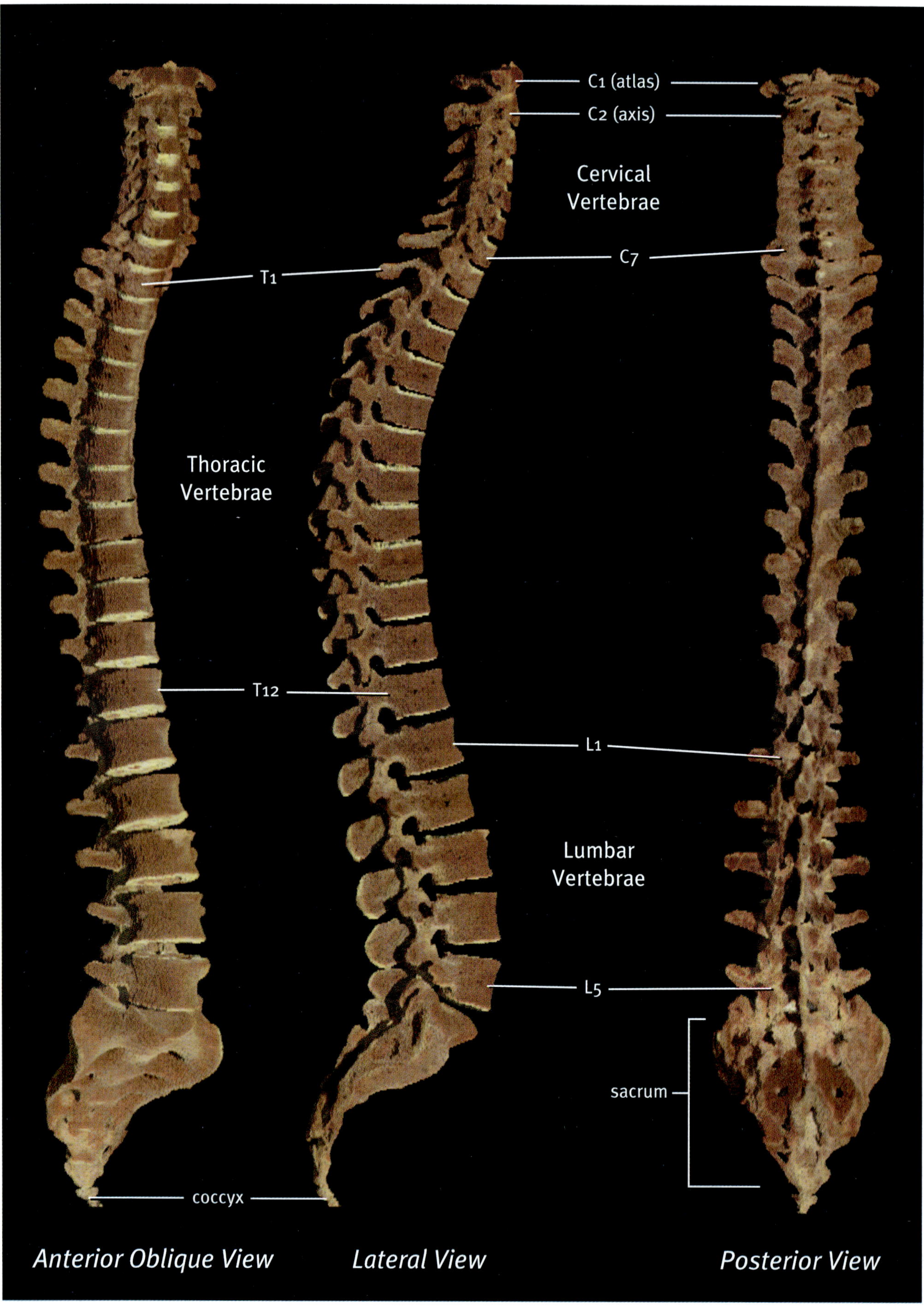

Spine and Pelvis

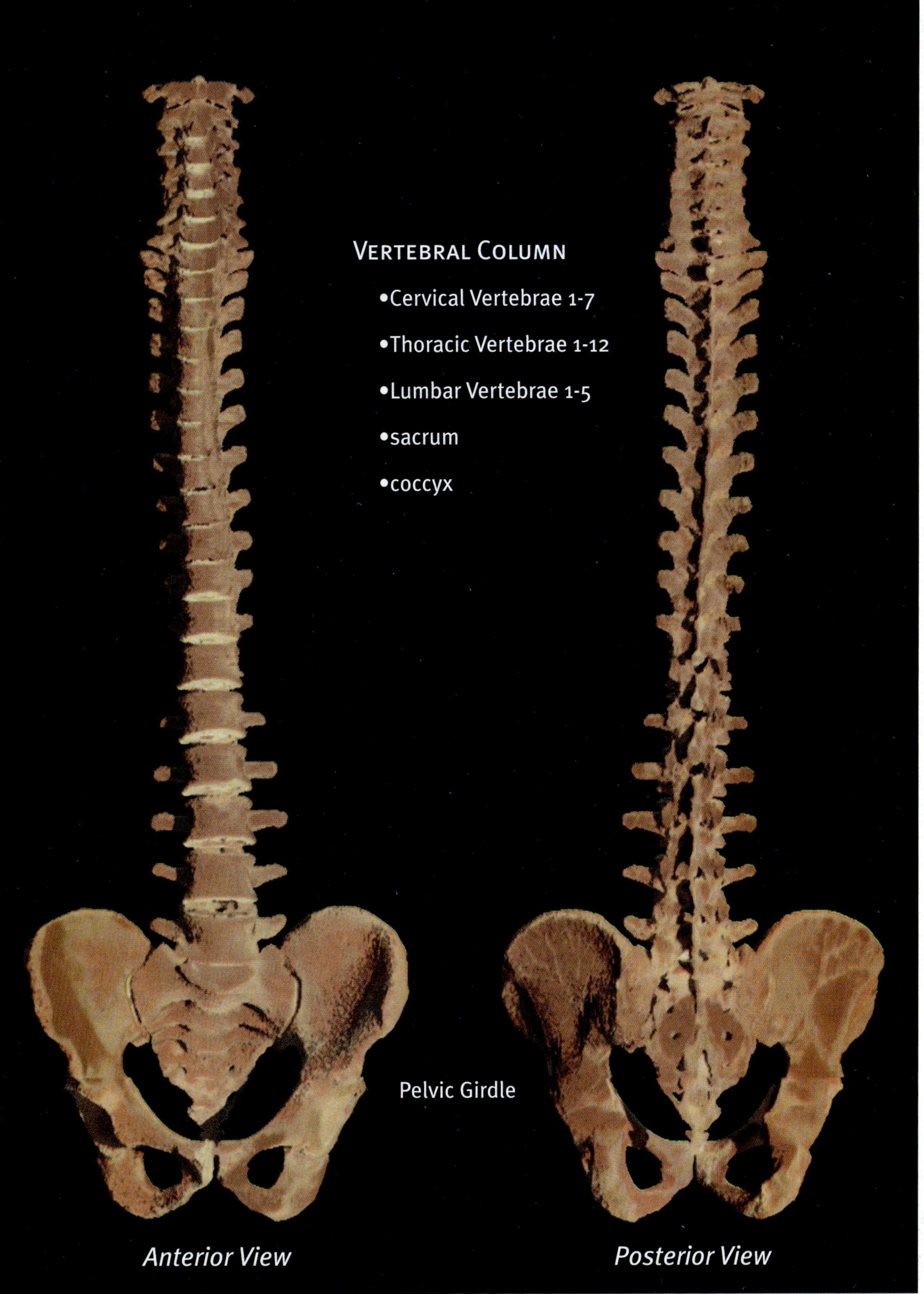

Atlas and Axis

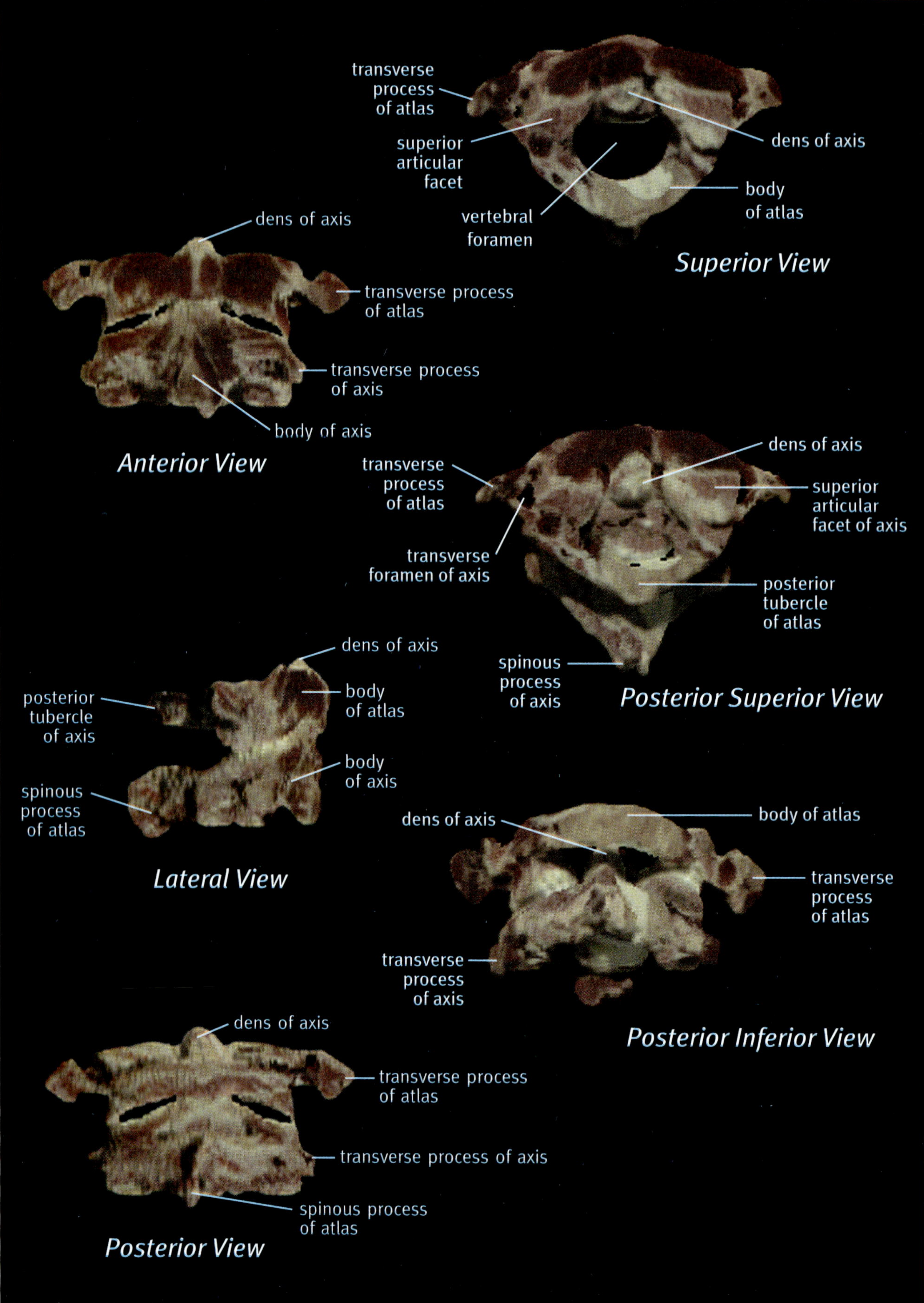

Cervical Vertebrae

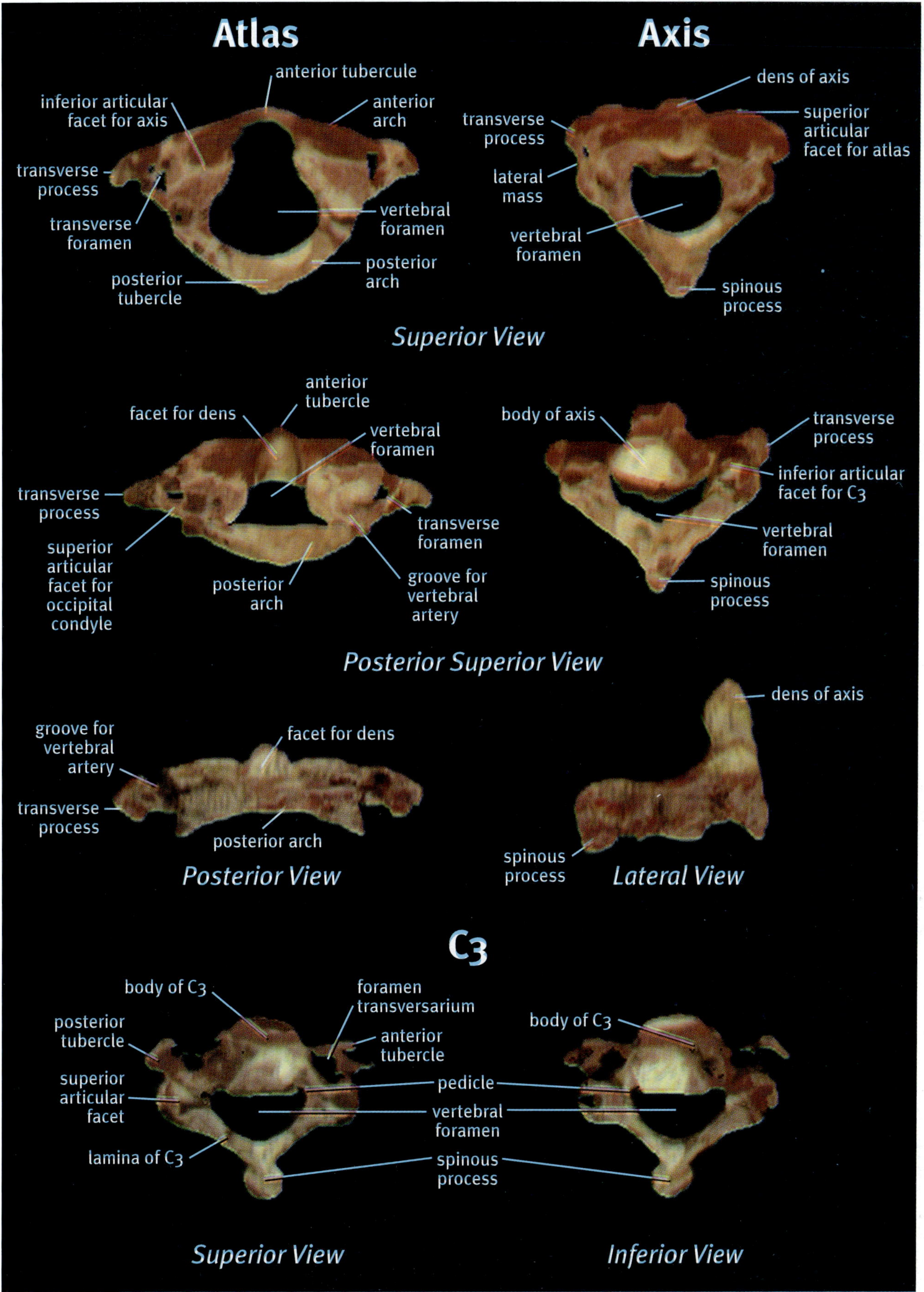

Thoracic Vertebrae

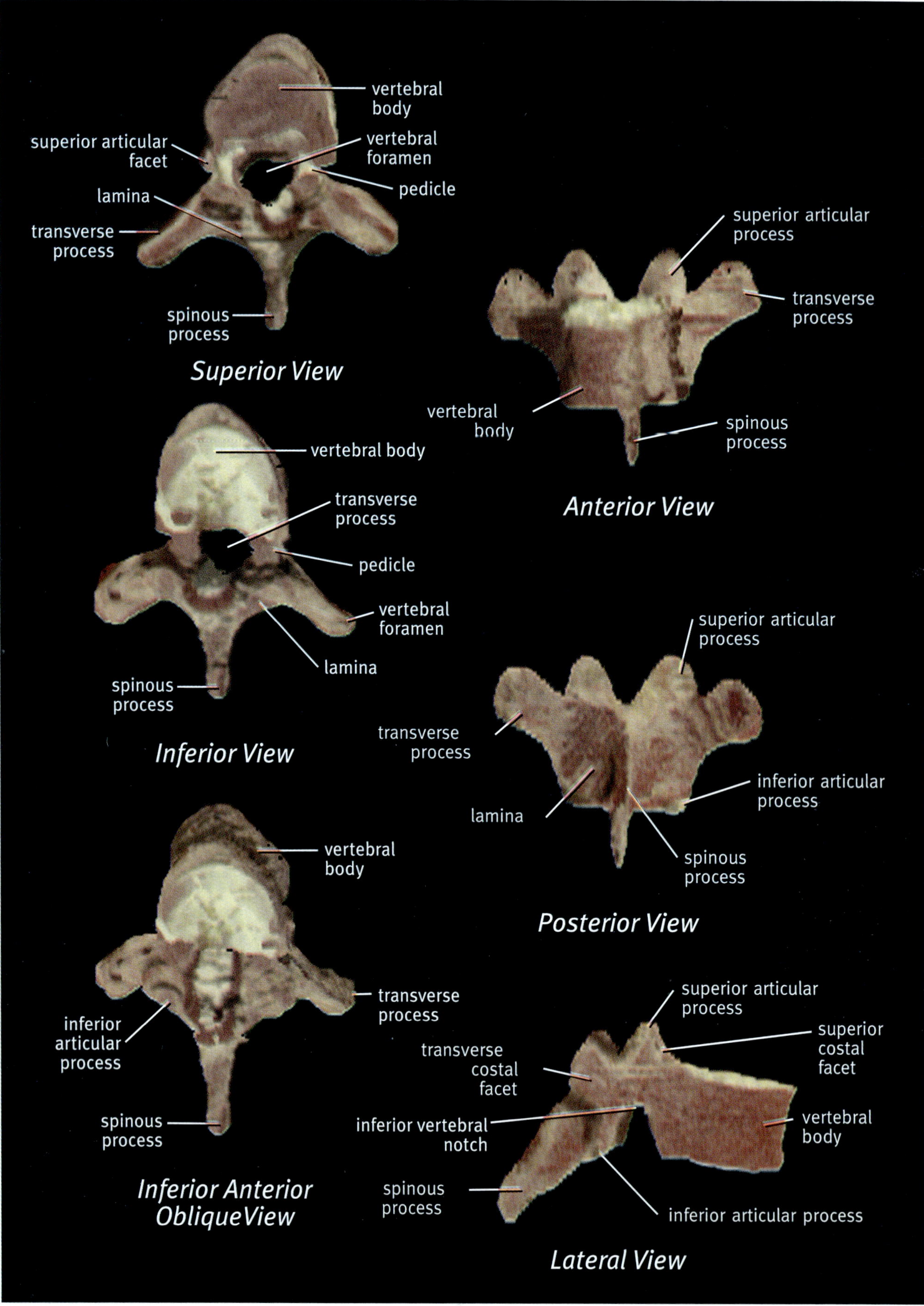

Lumbar Vertebrae

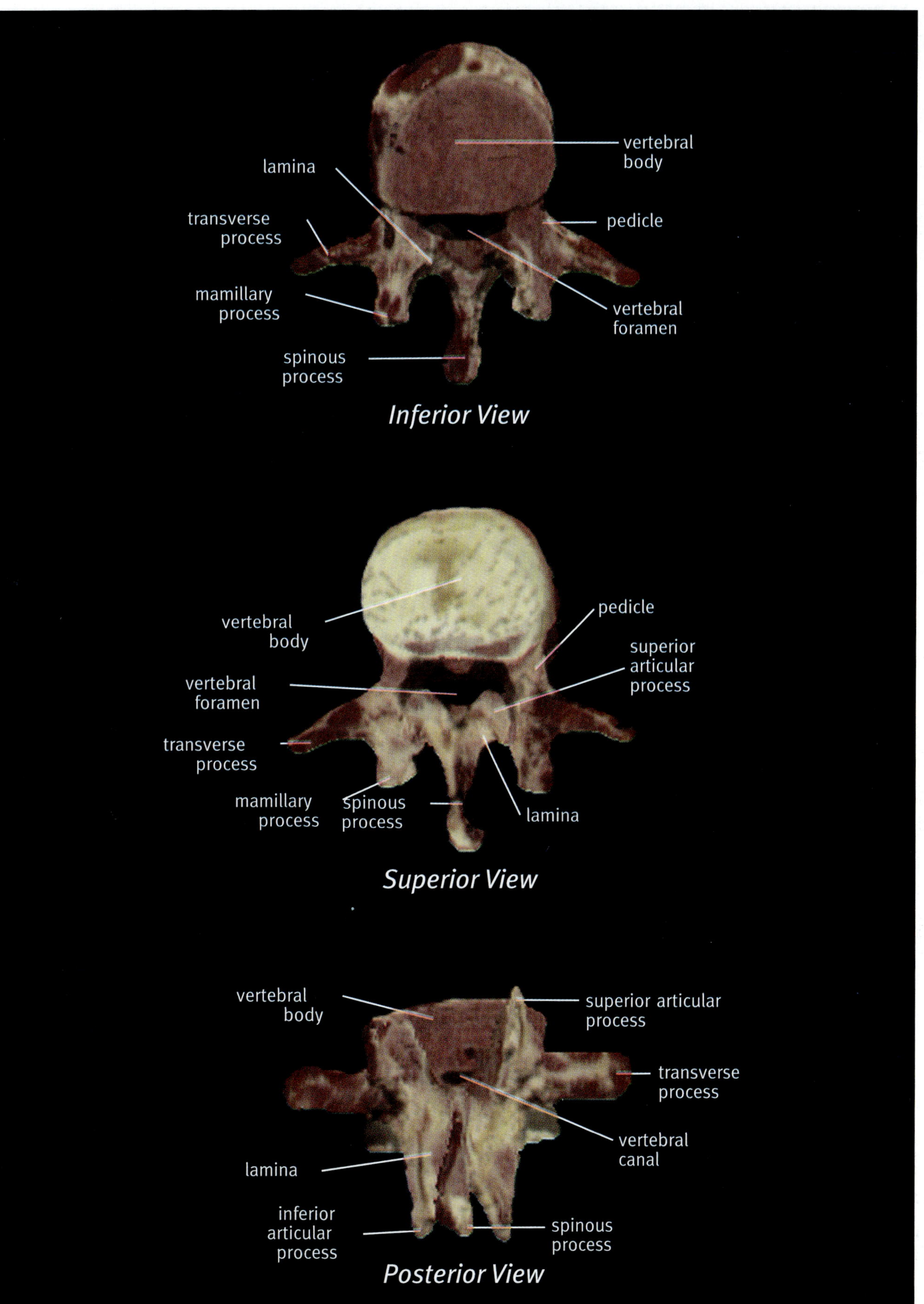

Inferior View

Superior View

Posterior View

Sacrum and Coccyx

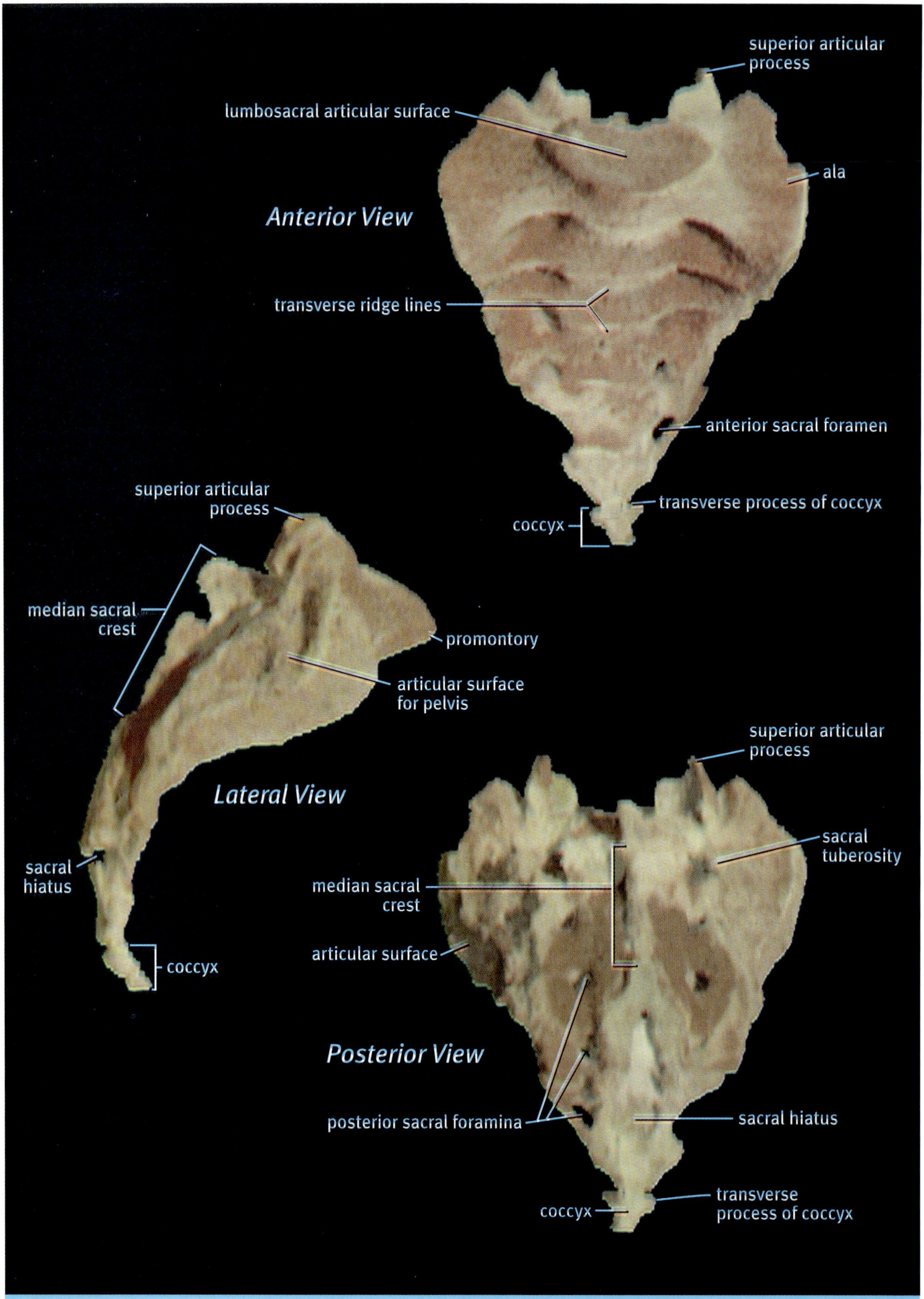

Ribcage

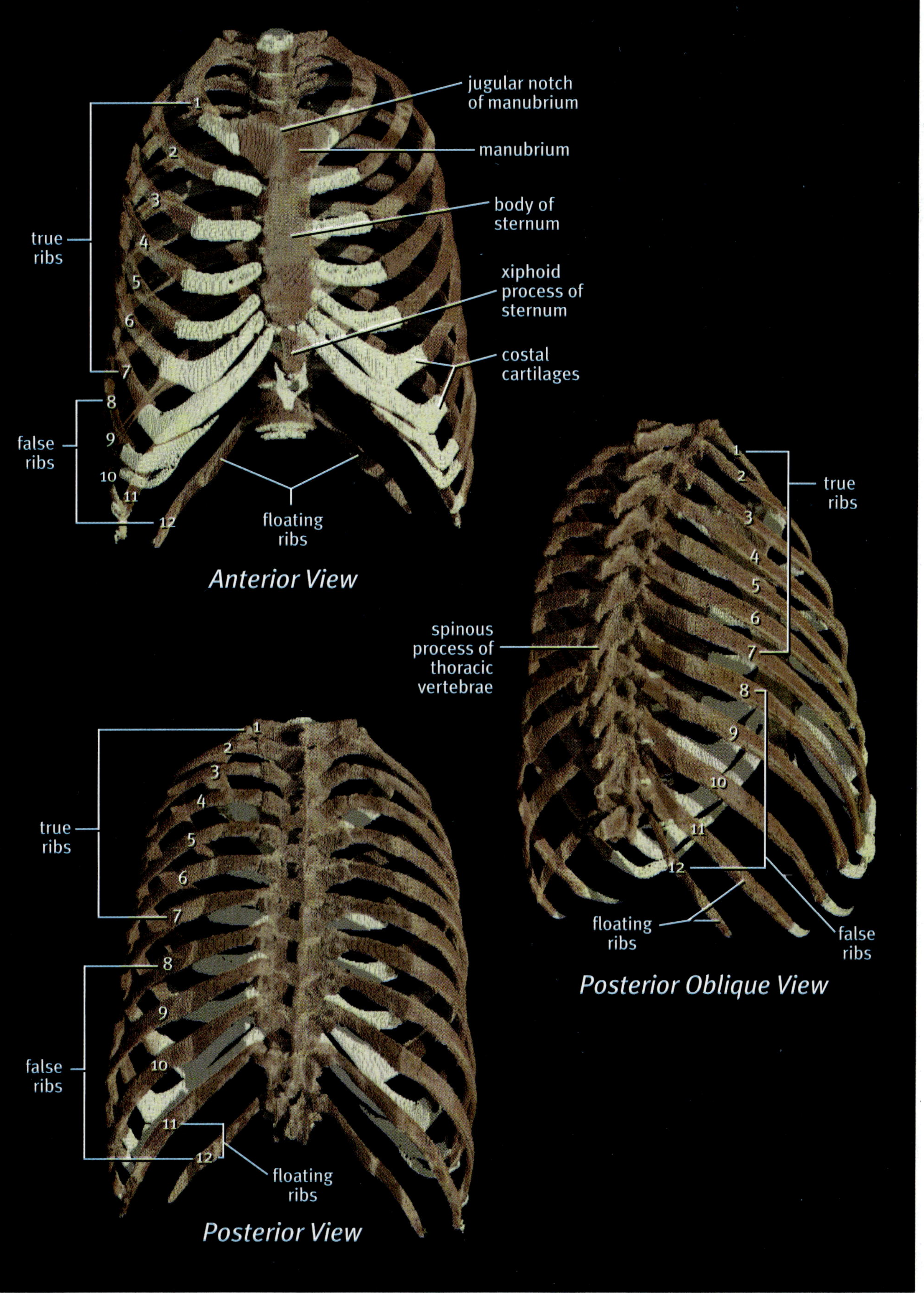

Anterior View

Posterior Oblique View

Posterior View

Sternum

jugular
notch
articular
surface
for clavicle
manubrium
angle
body
xiphoid
process
Anterior View
Posterior View
articular surface
for clavicle
jugular
notch
manubrium
manubrium
articular
surface
for first rib
body
xiphoid
process
Anterior Oblique
View
Lateral View

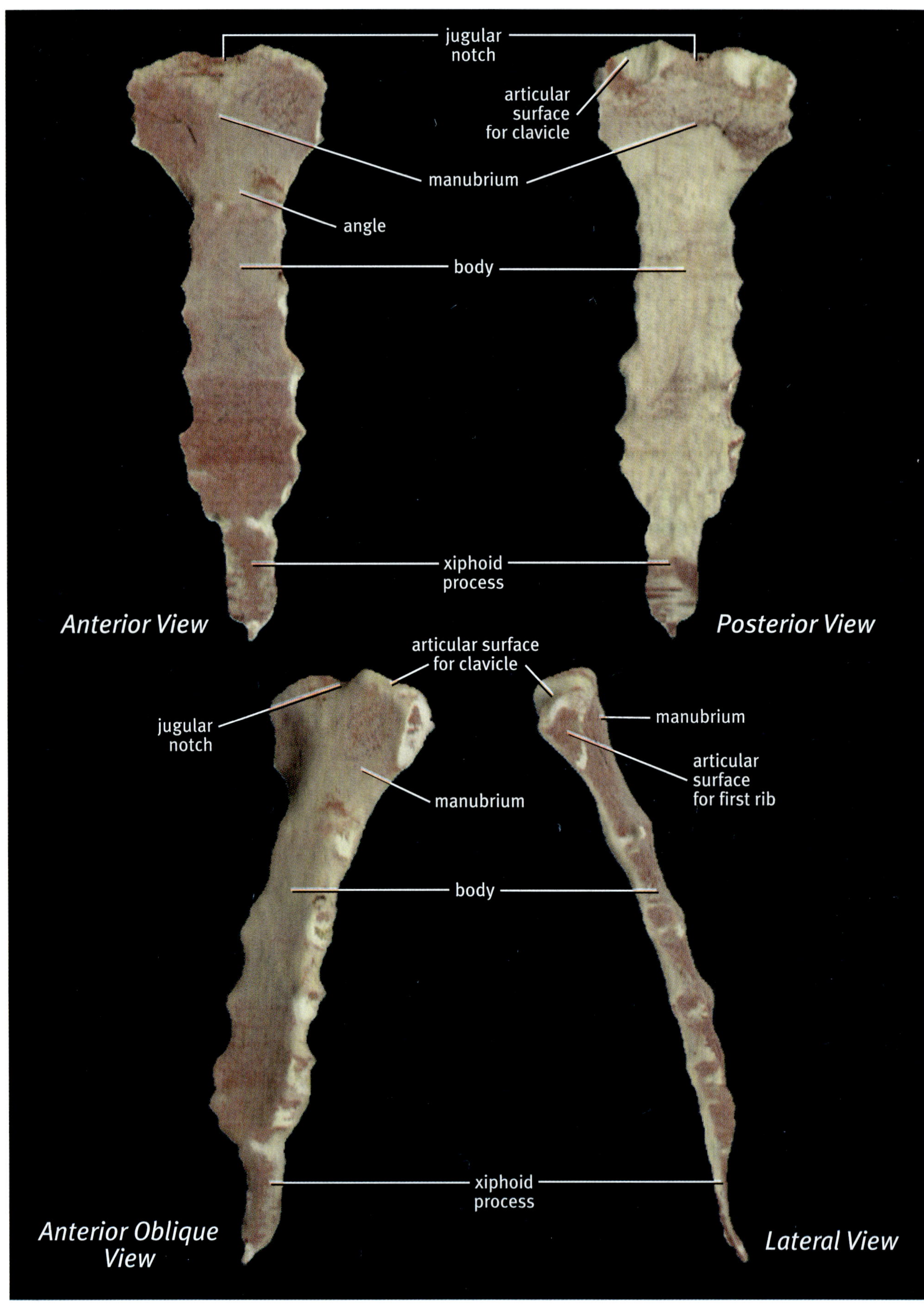

Rib

head
neck
articular facet
for vertebral body
Anterior View
neck
articular facet
for vertebral body
Anterior Oblique View
head
neck
facet for
transverse process
of thoracic vertebra
costal
groove
Posterior View
neck
head
angle
costal
groove
Anterior ObliqueView

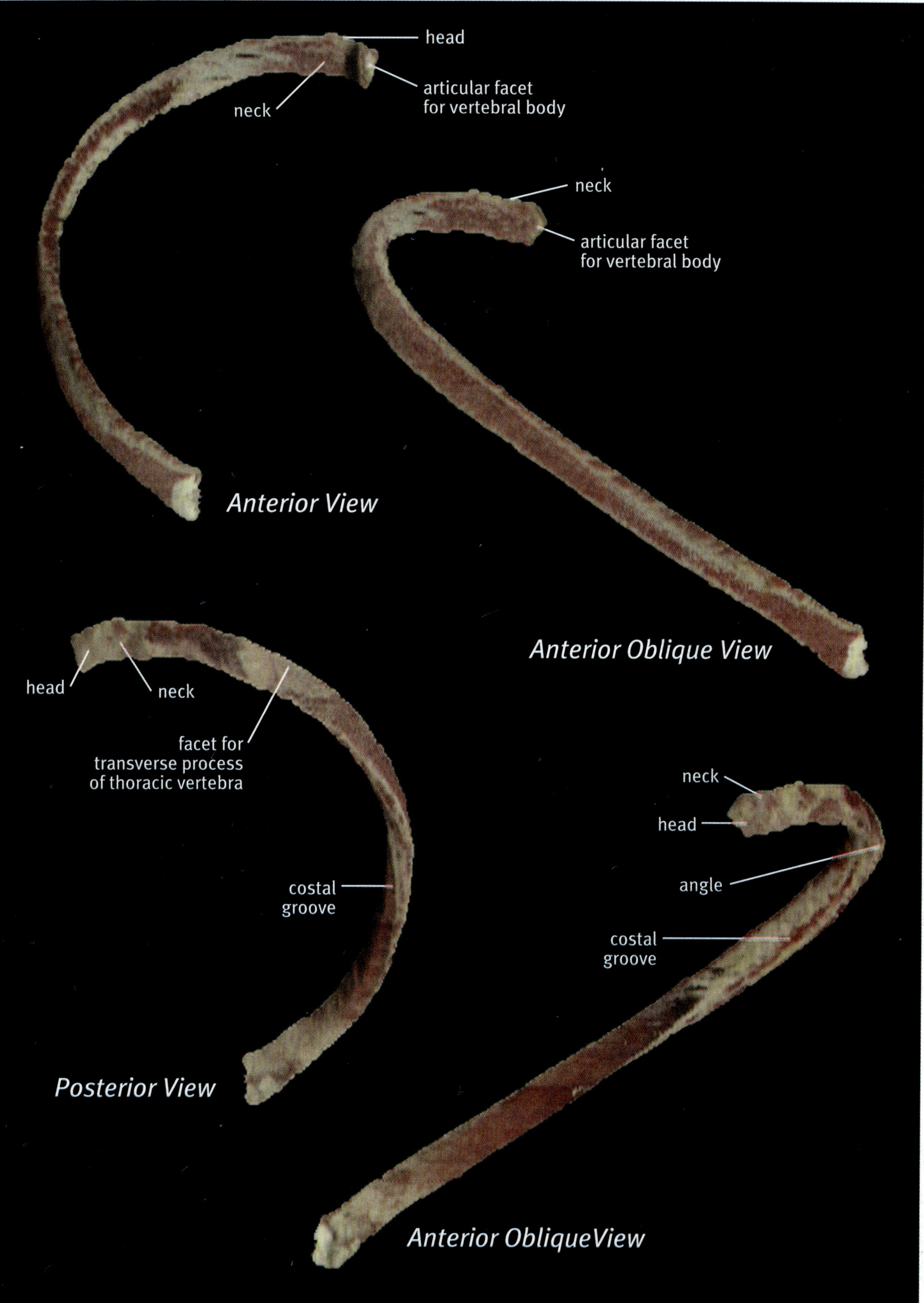

Upper Appendicular Skeleton 1

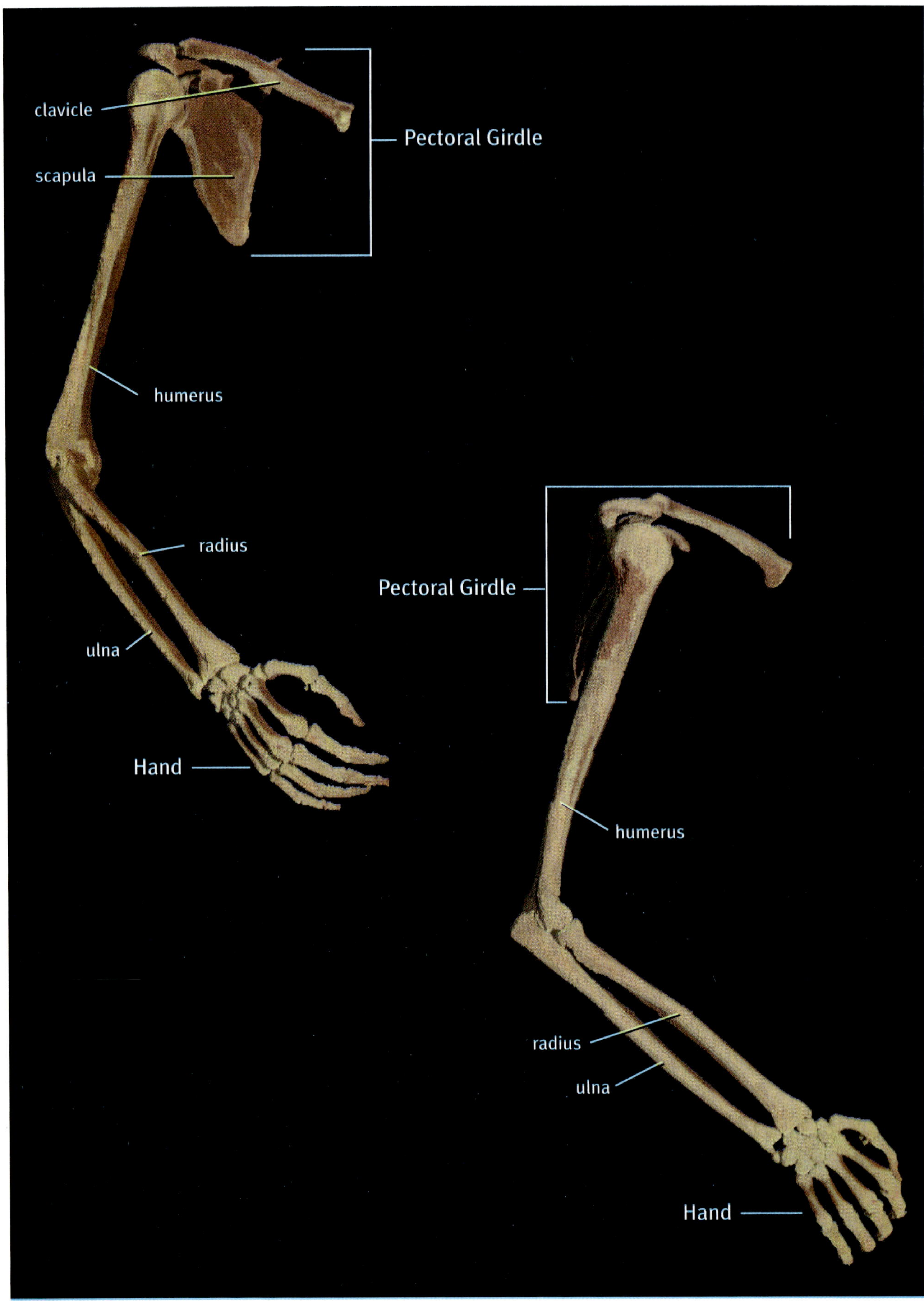

Upper Appendicular Skeleton 2

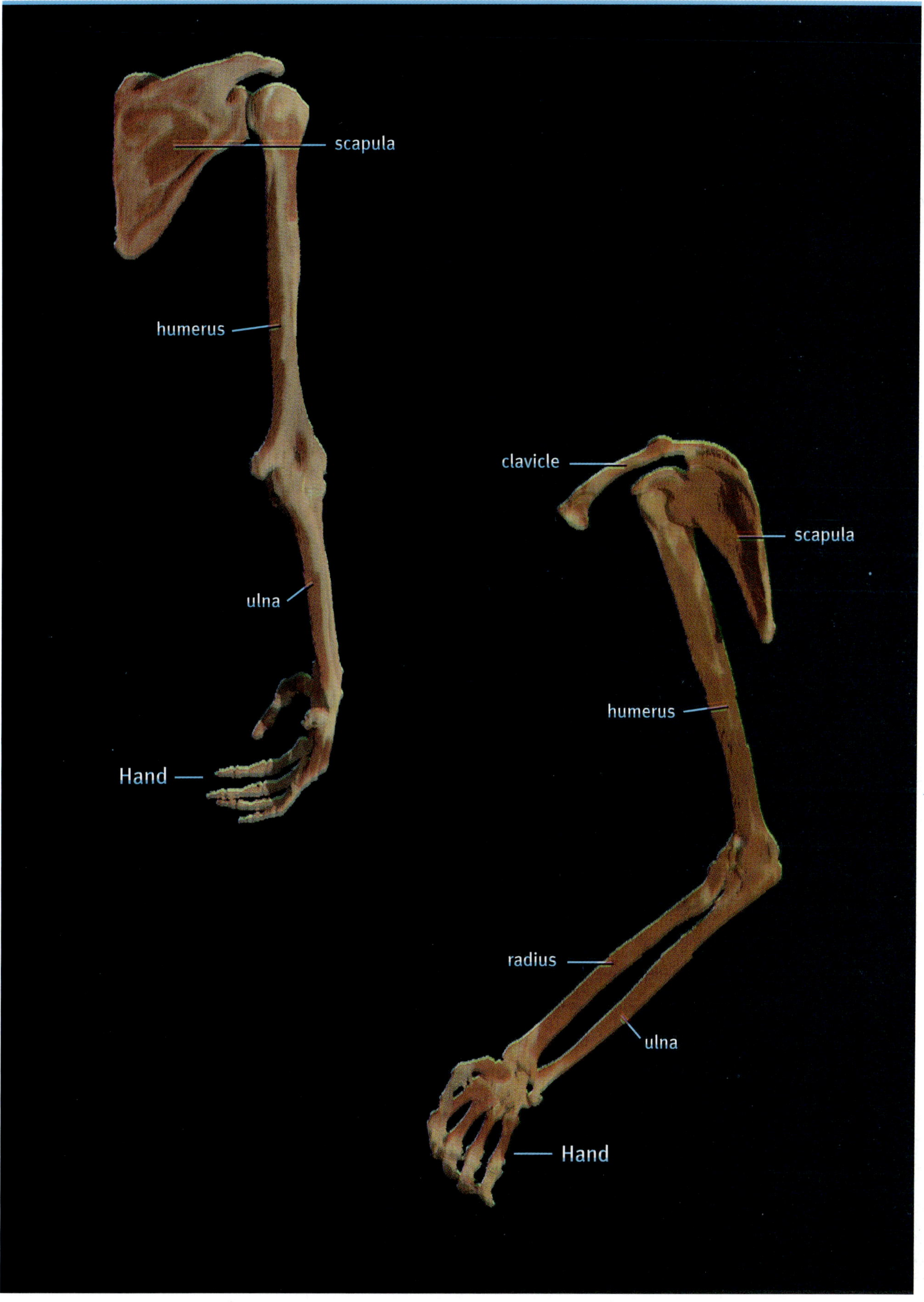

Pectoral Girdle 1

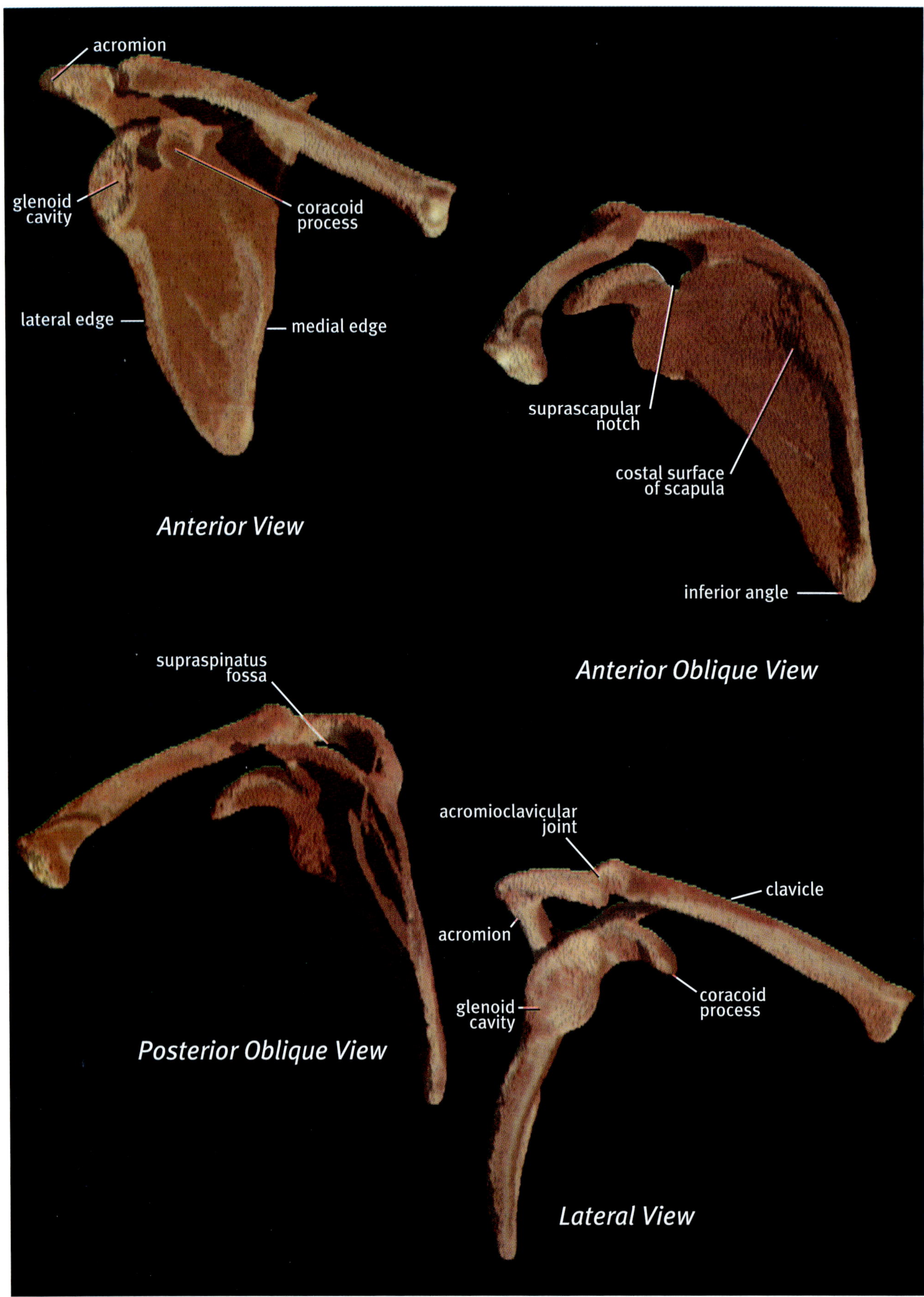

Pectoral Girdle 2

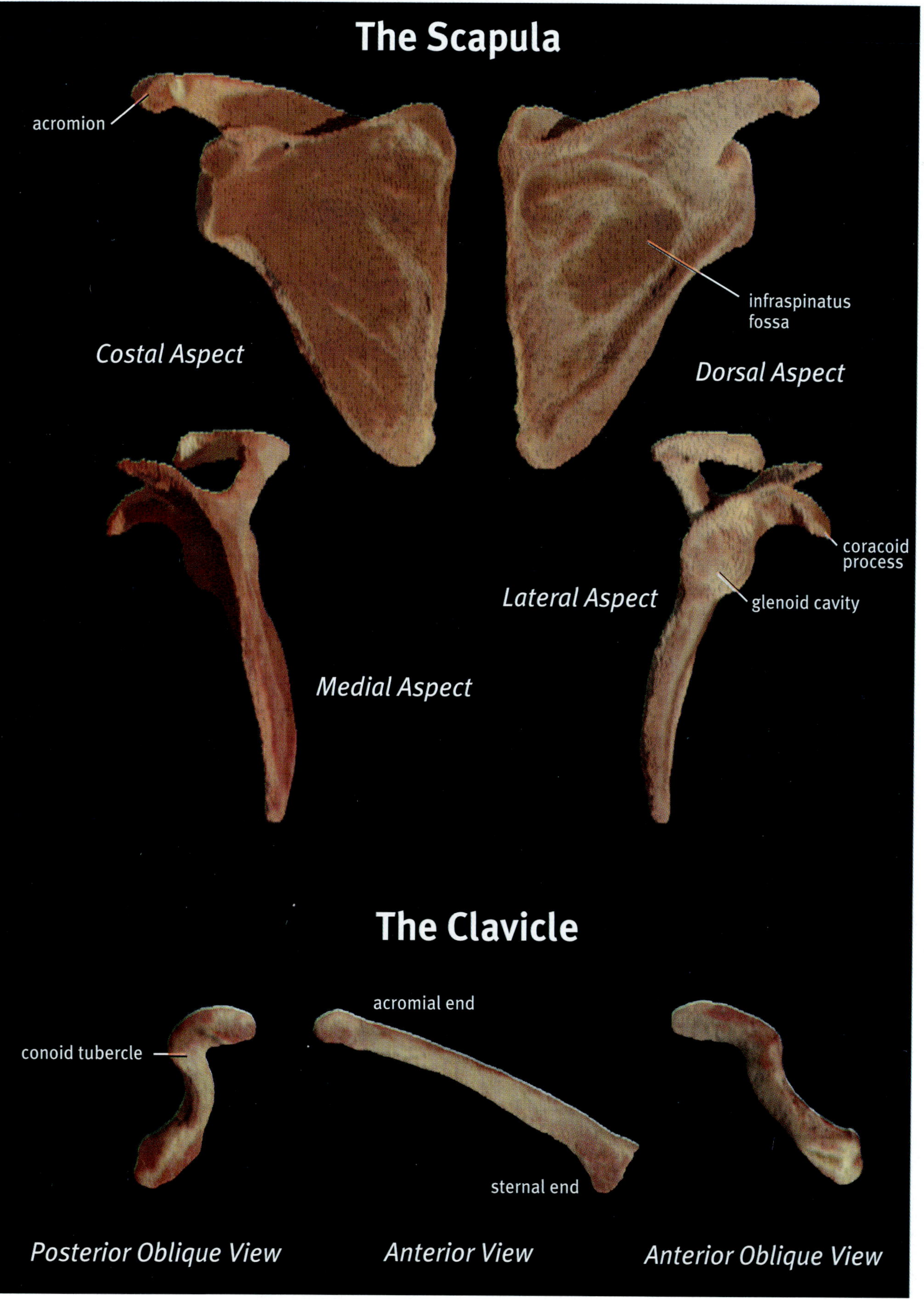

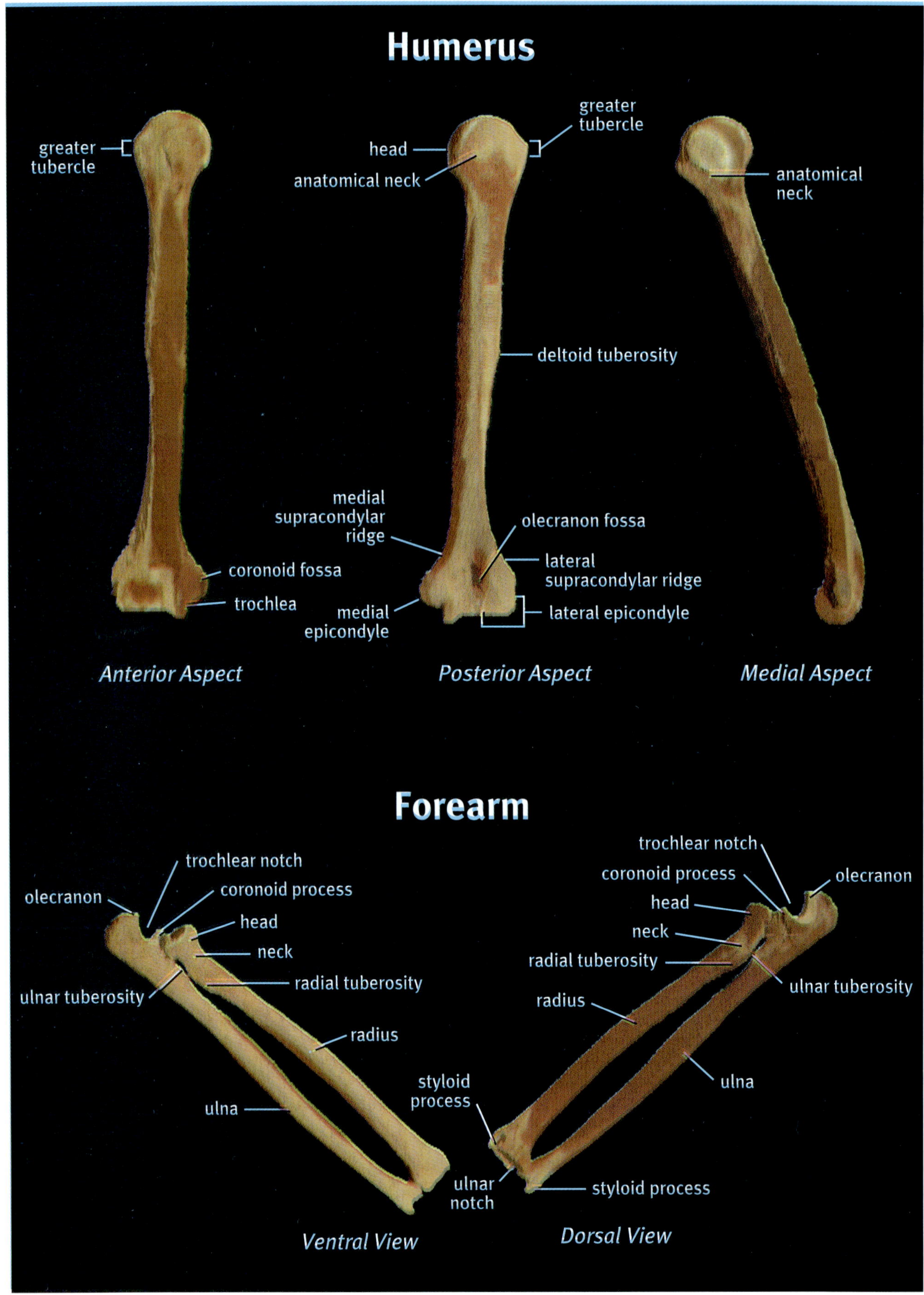
Humerus
greater tubercle
head
anatomical neck
greater tubercle
anatomical neck
deltoid tuberosity
medial supracondylar ridge
olecranon fossa
coronoid fossa
lateral supracondylar ridge
trochlea
medial epicondyle
lateral epicondyle
Anterior Aspect
Posterior Aspect
Medial Aspect
Forearm
trochlear notch
coronoid process
olecranon
head
neck
ulnar tuberosity
radial tuberosity
radius
ulna
styloid process
ulnar notch
trochlear notch
coronoid process
olecranon
head
neck
radial tuberosity
ulnar tuberosity
radius
ulna
styloid process
Ventral View
Dorsal View

Upper Extremity 2

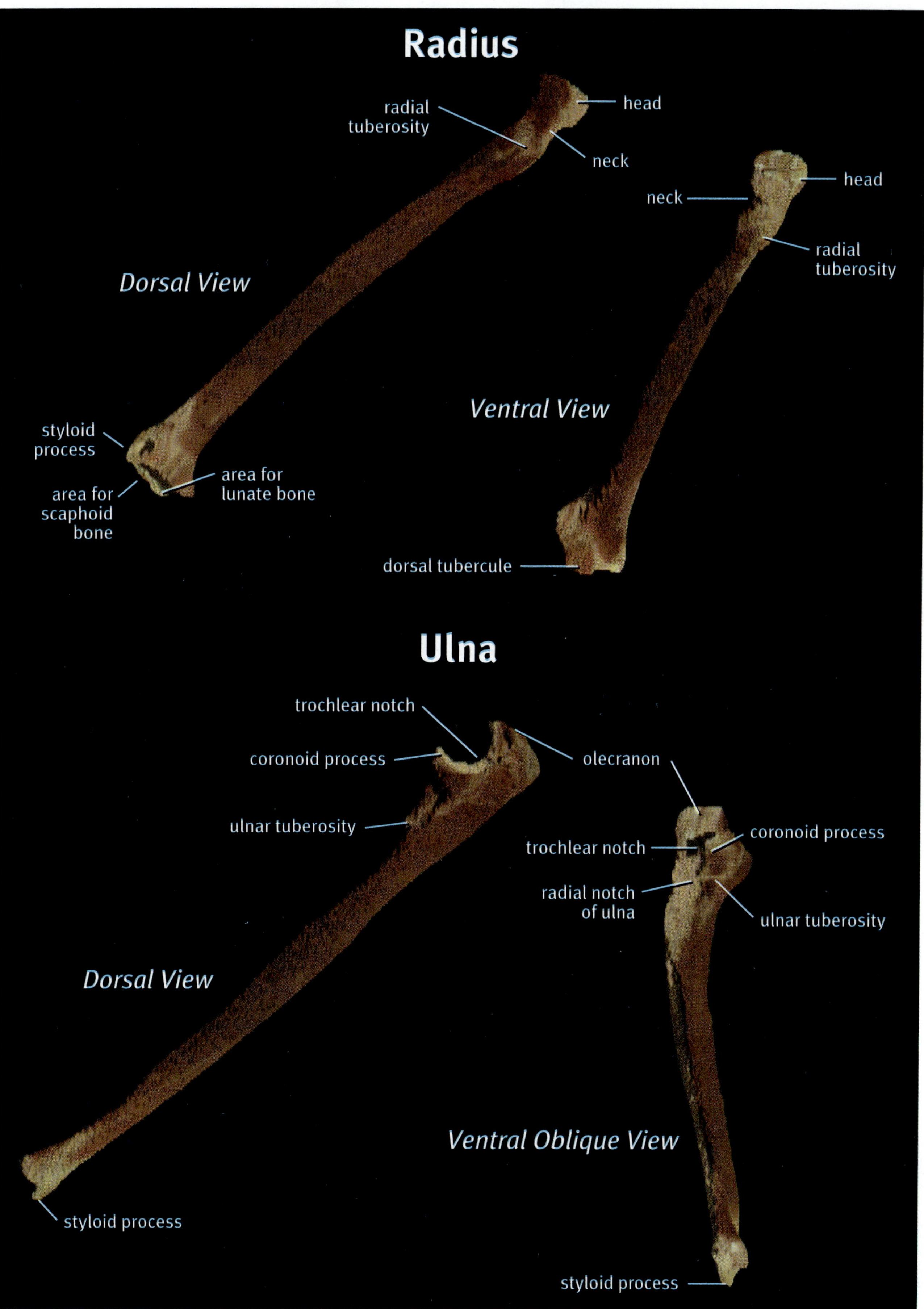

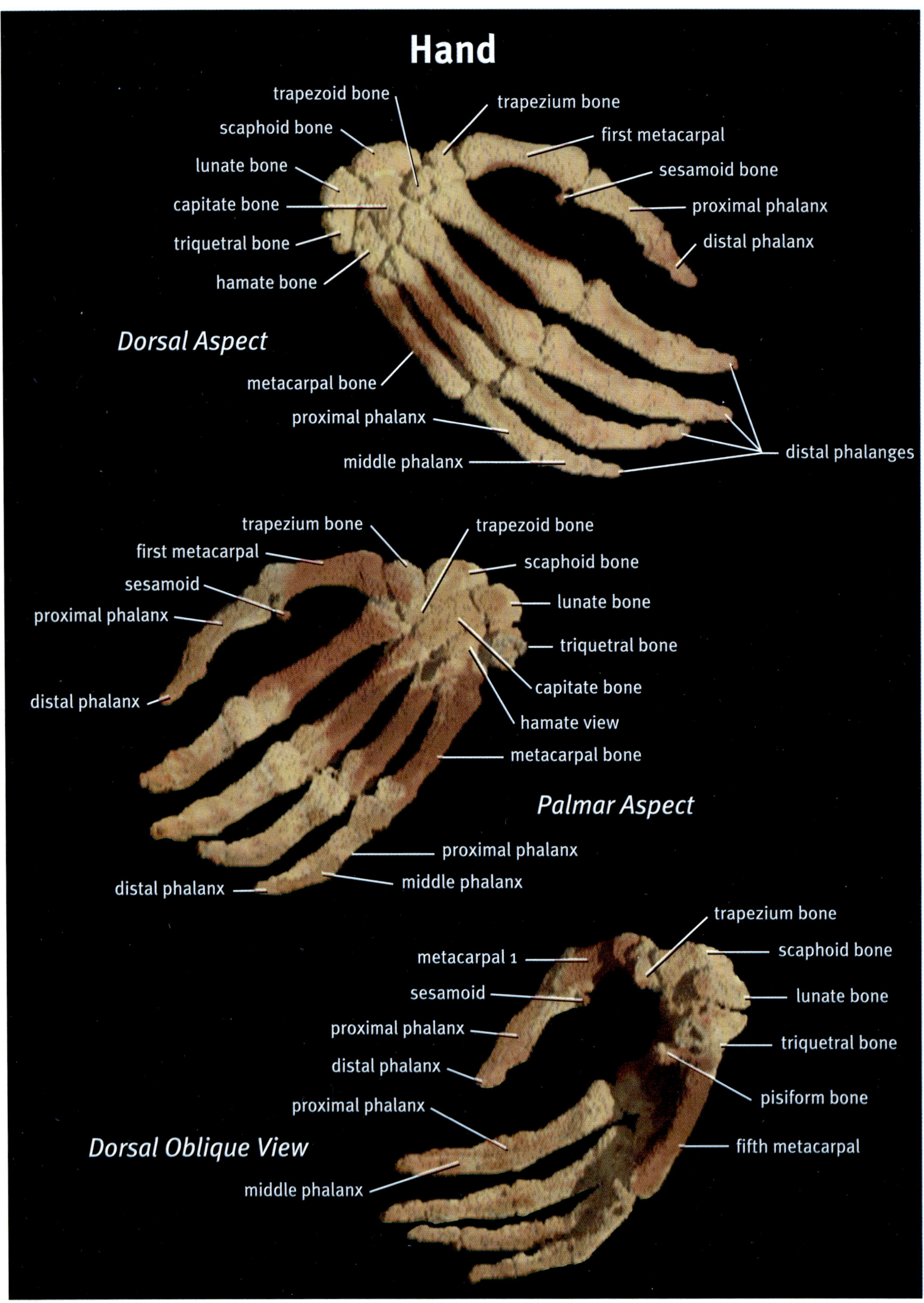
Hand
trapezoid bone
trapezium bone
scaphoid bone
first metacarpal
lunate bone
sesamoid bone
capitate bone
proximal phalanx
triquetral bone
distal phalanx
hamate bone
Dorsal Aspect
metacarpal bone
proximal phalanx
distal phalanges
middle phalanx
trapezium bone
trapezoid bone
first metacarpal
scaphoid bone
sesamoid
lunate bone
proximal phalanx
triquetral bone
capitate bone
distal phalanx
hamate view
metacarpal bone
Palmar Aspect
proximal phalanx
distal phalanx
middle phalanx
trapezium bone
metacarpal 1
scaphoid bone
sesamoid
lunate bone
proximal phalanx
triquetral bone
distal phalanx
pisiform bone
proximal phalanx
Dorsal Oblique View
fifth metacarpal
middle phalanx

Lower Appendicular Skeleton

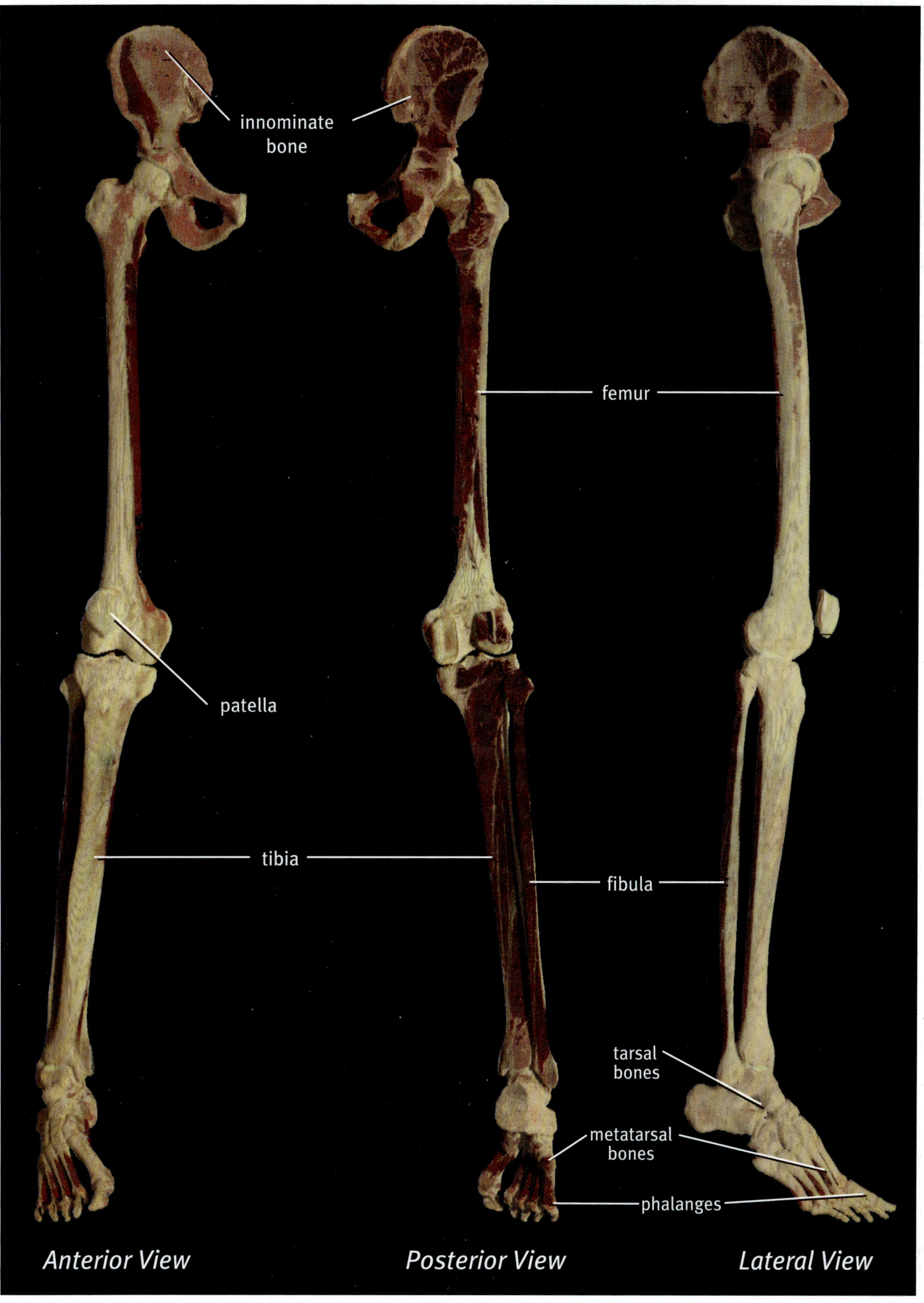

Pelvis

superior articular process of sacrum
lumbosacral articular surface
iliac crest
wing of ilium
iliopubic eminence
acetabulum
obturator foramen
ischiatic tuberosity
pubic tubercle
coccyx
pubic symphysis
Anterior View
posterior superior iliac spine
wing of ilium
superior articular process
sacral tuberosity
median sacral crest
posterior sacral foramen
greater sciatic notch
coccyx
obturator foramen
superior pubic ramus
Posterior View
pubic symphysis
wing of ilium
iliac crest
superior articular process of sacrum
ala
lumbosacral articular surface of sacrum
coccyx
superior pubic ramus
pubic symphysis
Superior View
ischiatic tuberosity
ischial spine
wing of ilium
superior articular process of sacrum
Inferior View

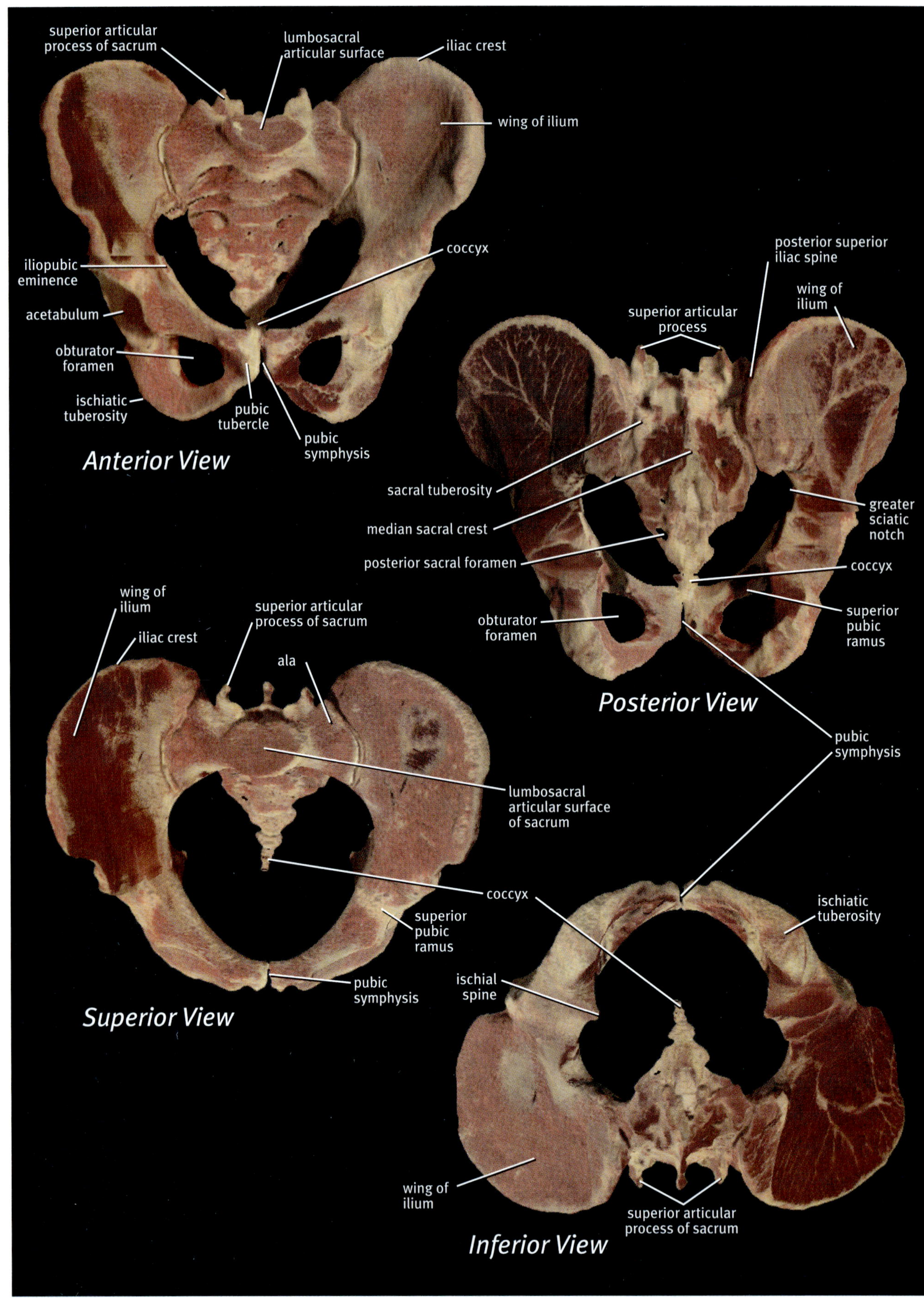

Innominate Bone, Sacrum and Coccyx

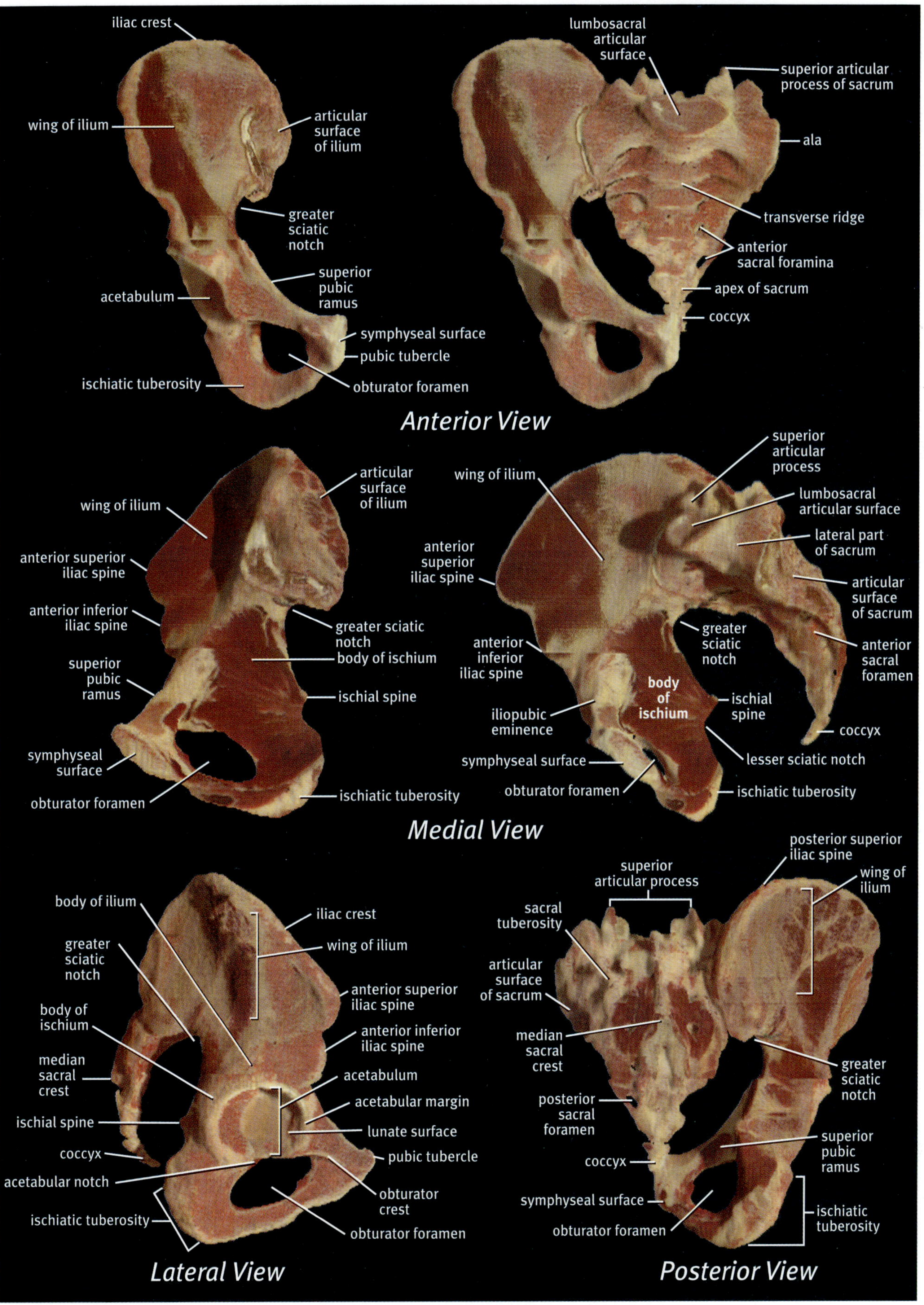

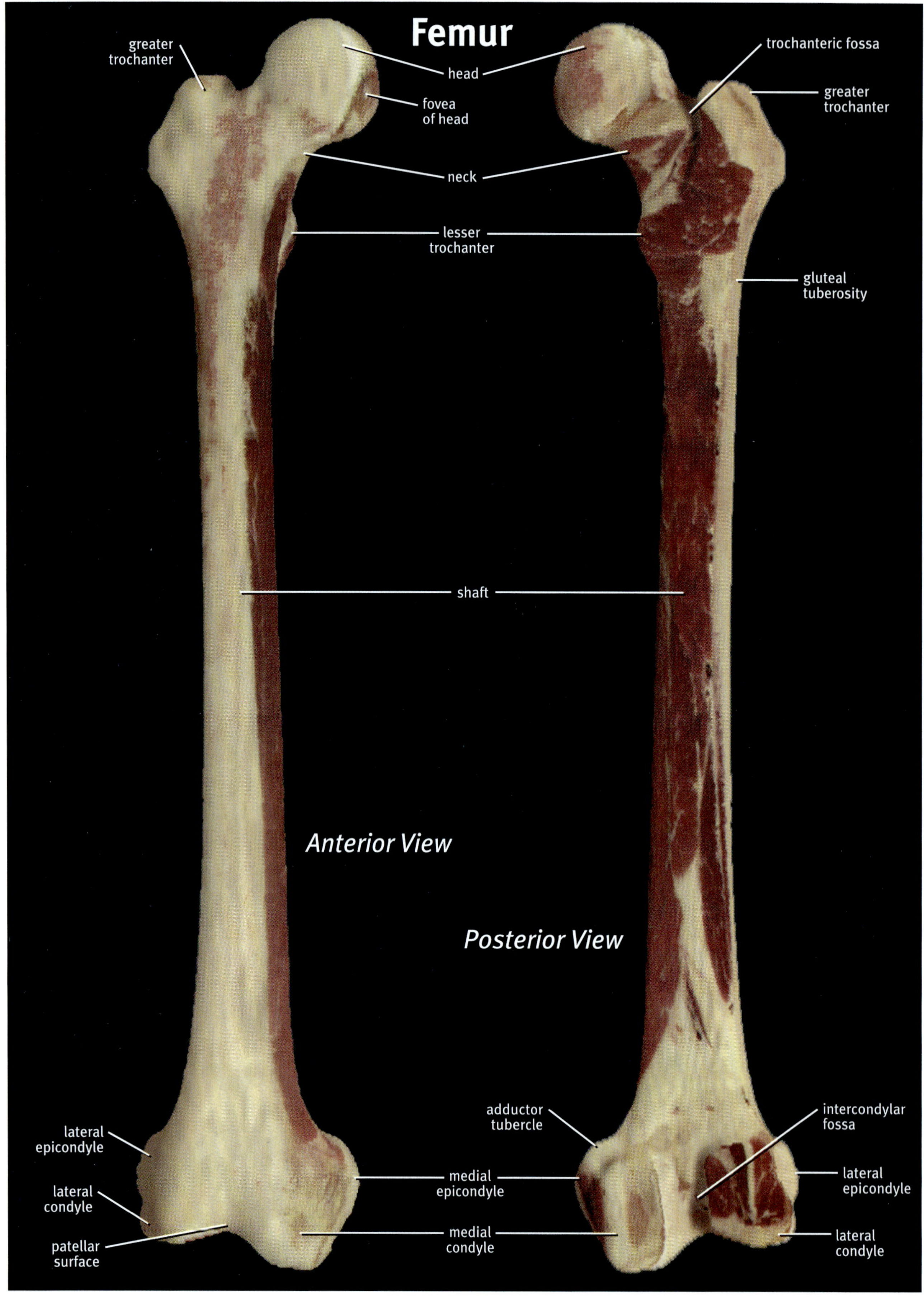
Femur
greater trochanter
head
fovea of head
trochanteric fossa
greater trochanter
neck
lesser trochanter
gluteal tuberosity
shaft
Anterior View
Posterior View
lateral epicondyle
lateral condyle
patellar surface
adductor tubercle
medial epicondyle
medial condyle
intercondylar fossa
lateral epicondyle
lateral condyle

Lower Extremity 2

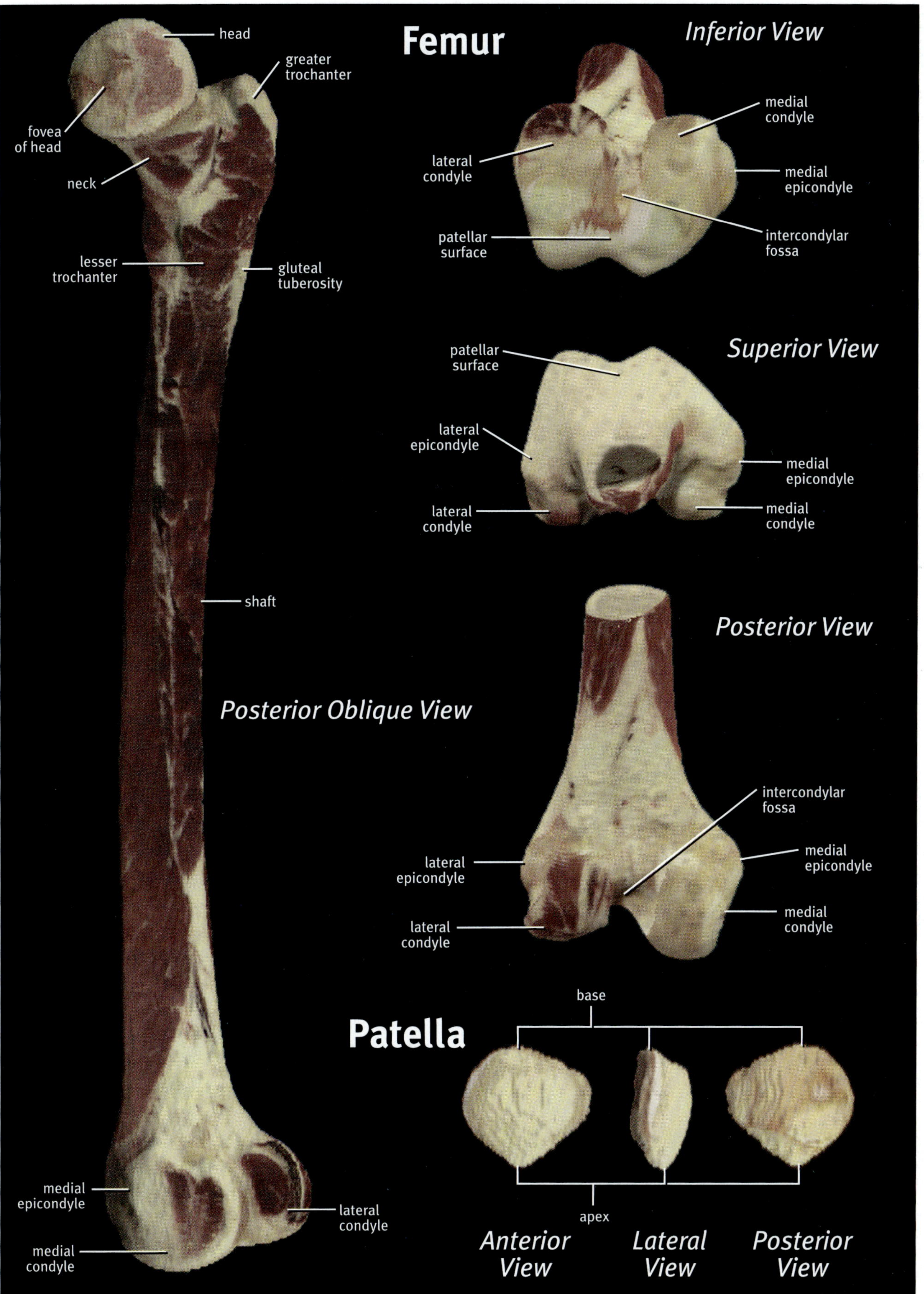

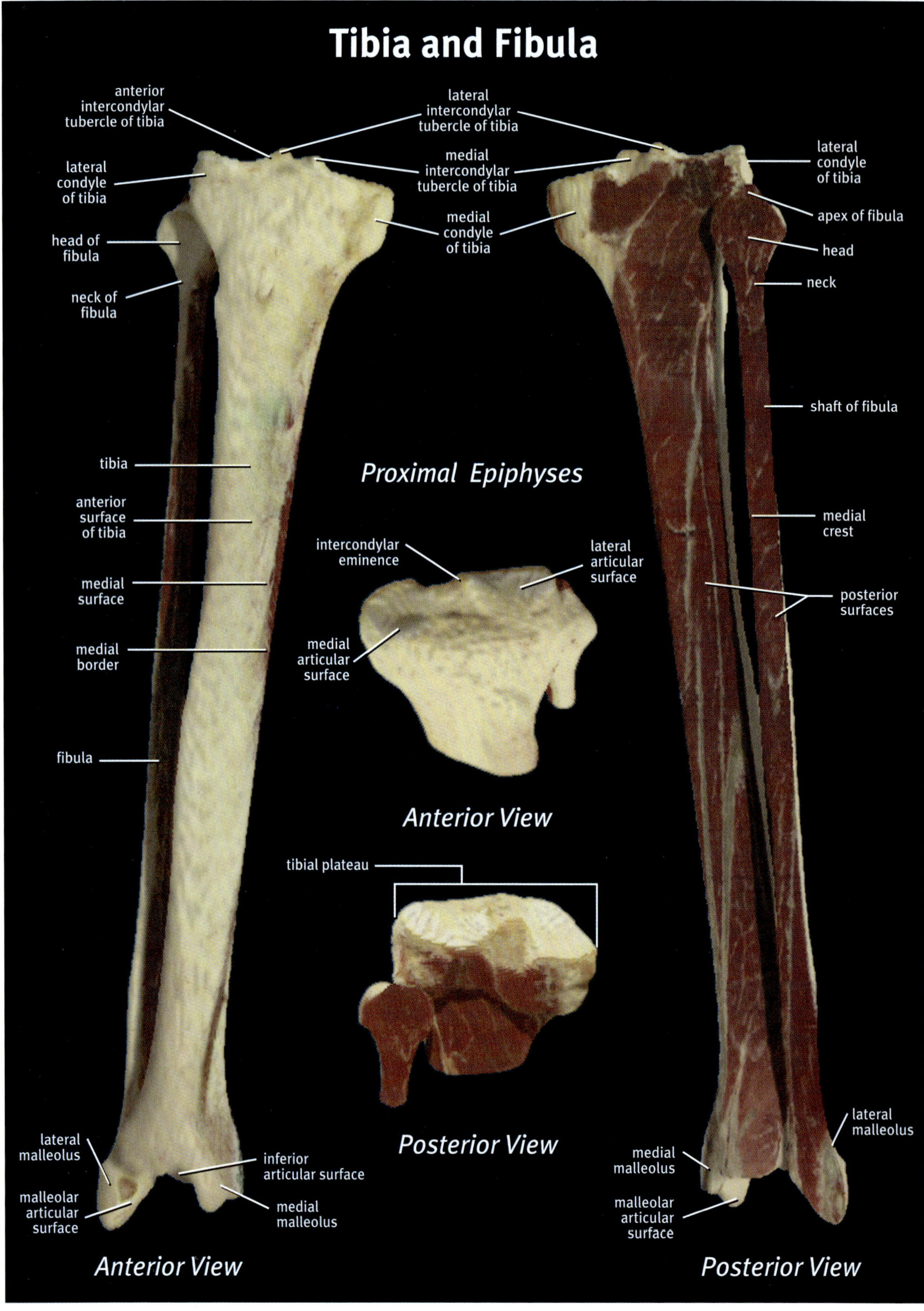
Tibia and Fibula
anterior intercondylar tubercle of tibia
lateral intercondylar tubercle of tibia
lateral condyle of tibia
medial intercondylar tubercle of tibia
lateral condyle of tibia
head of fibula
medial condyle of tibia
apex of fibula
head
neck of fibula
neck
shaft of fibula
tibia
Proximal Epiphyses
anterior surface of tibia
medial crest
intercondylar eminence
lateral articular surface
medial surface
posterior surfaces
medial articular surface
medial border
fibula
Anterior View
tibial plateau
lateral malleolus
medial malleolus
lateral malleolus
inferior articular surface
Posterior View
malleolar articular surface
medial malleolus
malleolar articular surface
Anterior View
Posterior View

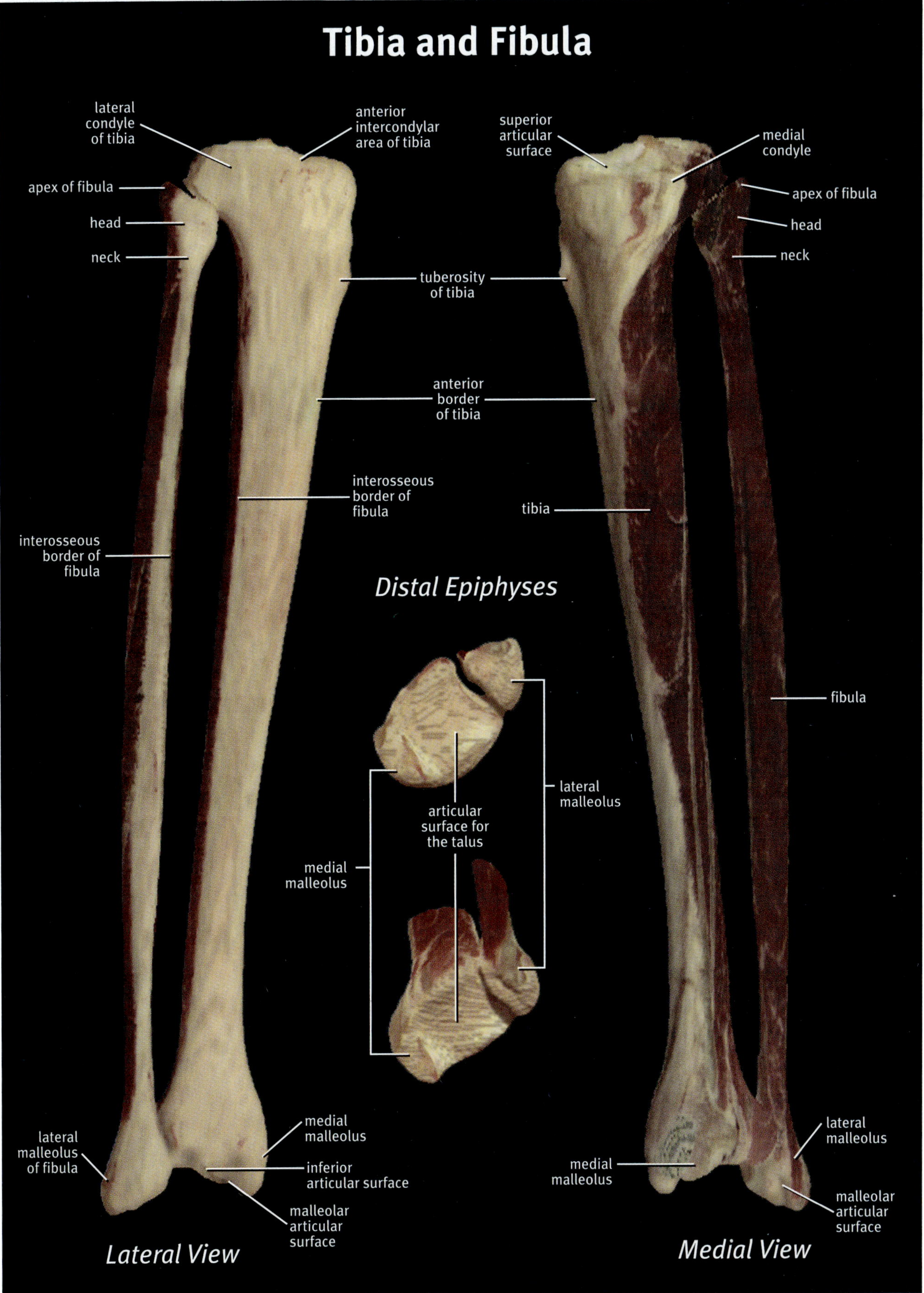
Tibia and Fibula
lateral condyle of tibia
anterior intercondylar area of tibia
superior articular surface
medial condyle
apex of fibula
head
neck
apex of fibula
head
neck
tuberosity of tibia
anterior border of tibia
interosseous border of fibula
tibia
interosseous border of fibula
Distal Epiphyses
fibula
lateral malleolus
articular surface for the talus
medial malleolus
lateral malleolus of fibula
medial malleolus
inferior articular surface
malleolar articular surface
lateral malleolus
medial malleolus
malleolar articular surface
Lateral View
Medial View

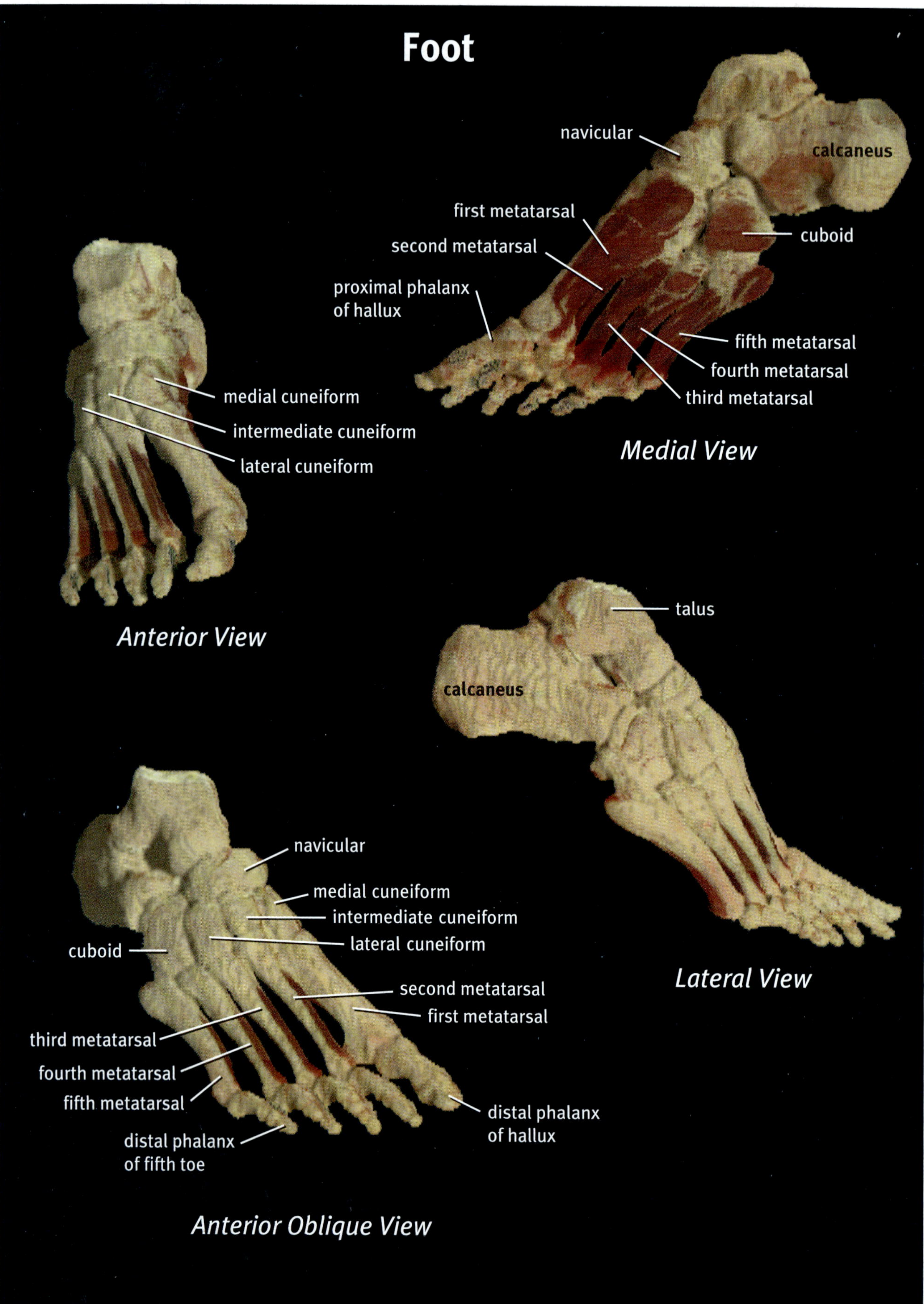
Foot
navicular
calcaneus
first metatarsal
cuboid
second metatarsal
proximal phalanx
of hallux
fifth metatarsal
fourth metatarsal
third metatarsal
Medial View
medial cuneiform
intermediate cuneiform
lateral cuneiform
Anterior View
talus
calcaneus
Lateral View
navicular
medial cuneiform
intermediate cuneiform
lateral cuneiform
cuboid
second metatarsal
first metatarsal
third metatarsal
fourth metatarsal
fifth metatarsal
distal phalanx
of hallux
distal phalanx
of fifth toe
Anterior Oblique View

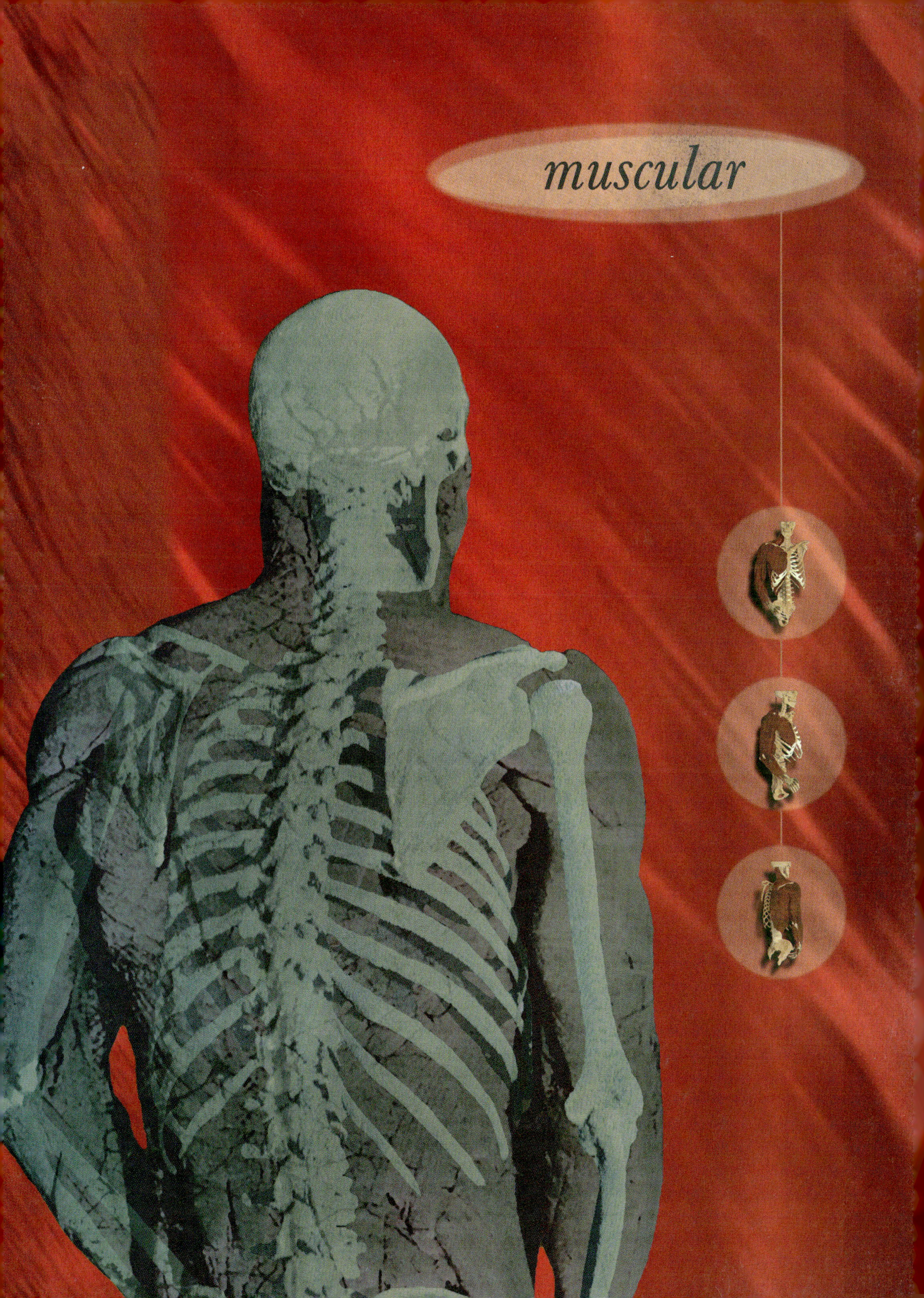
muscular

Muscular System 1

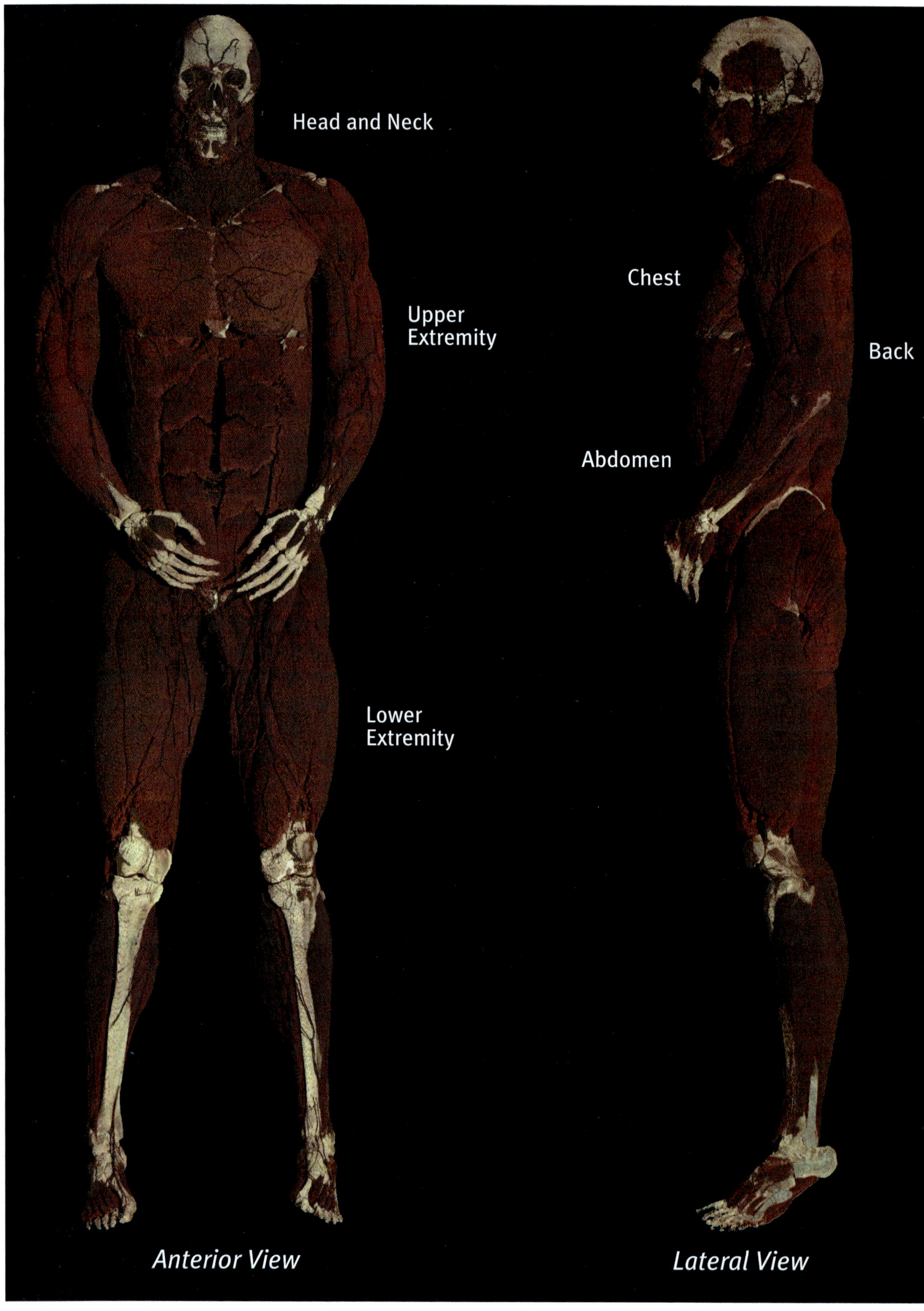

Muscular System 2

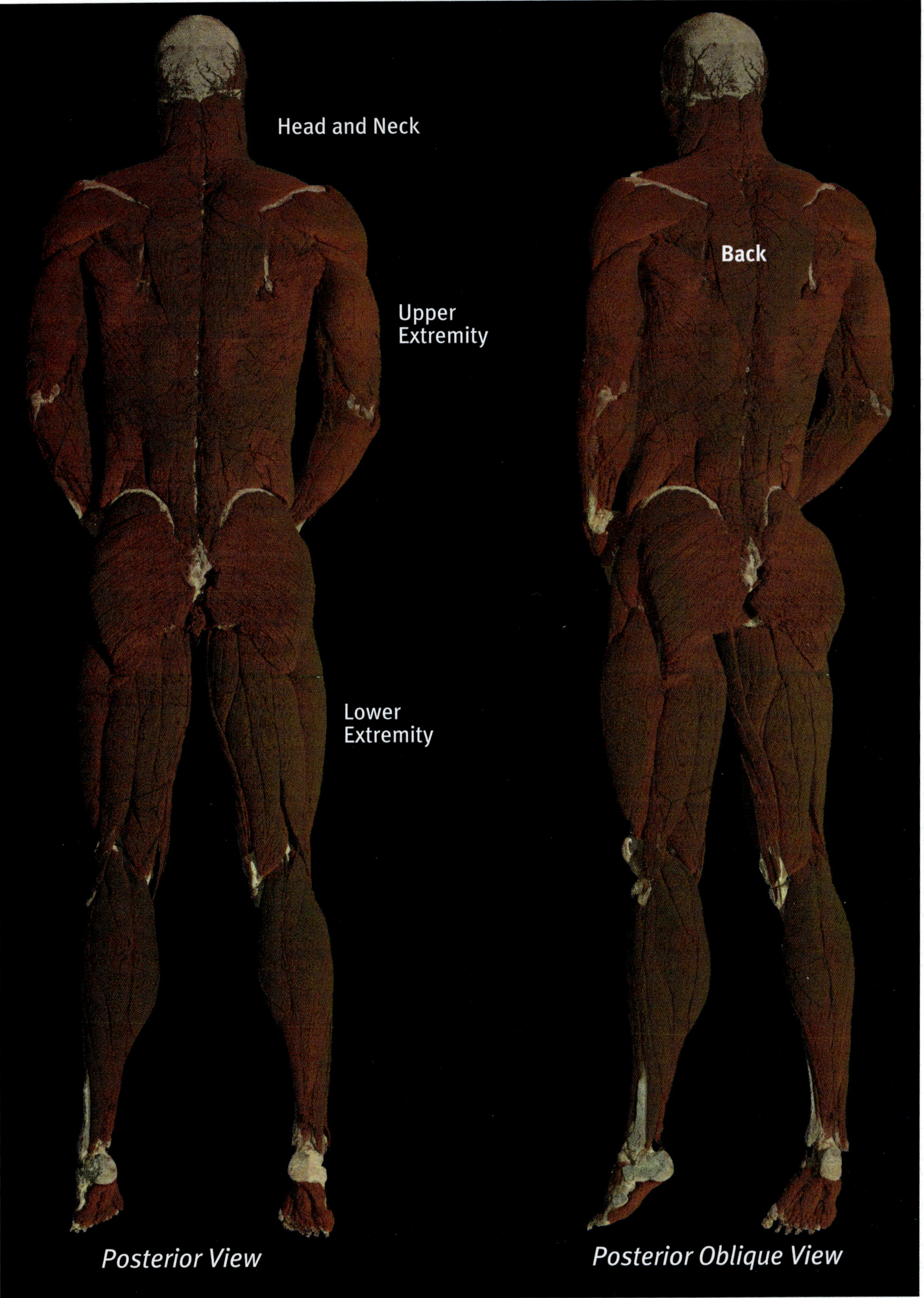

Muscles of the Head and Neck 1

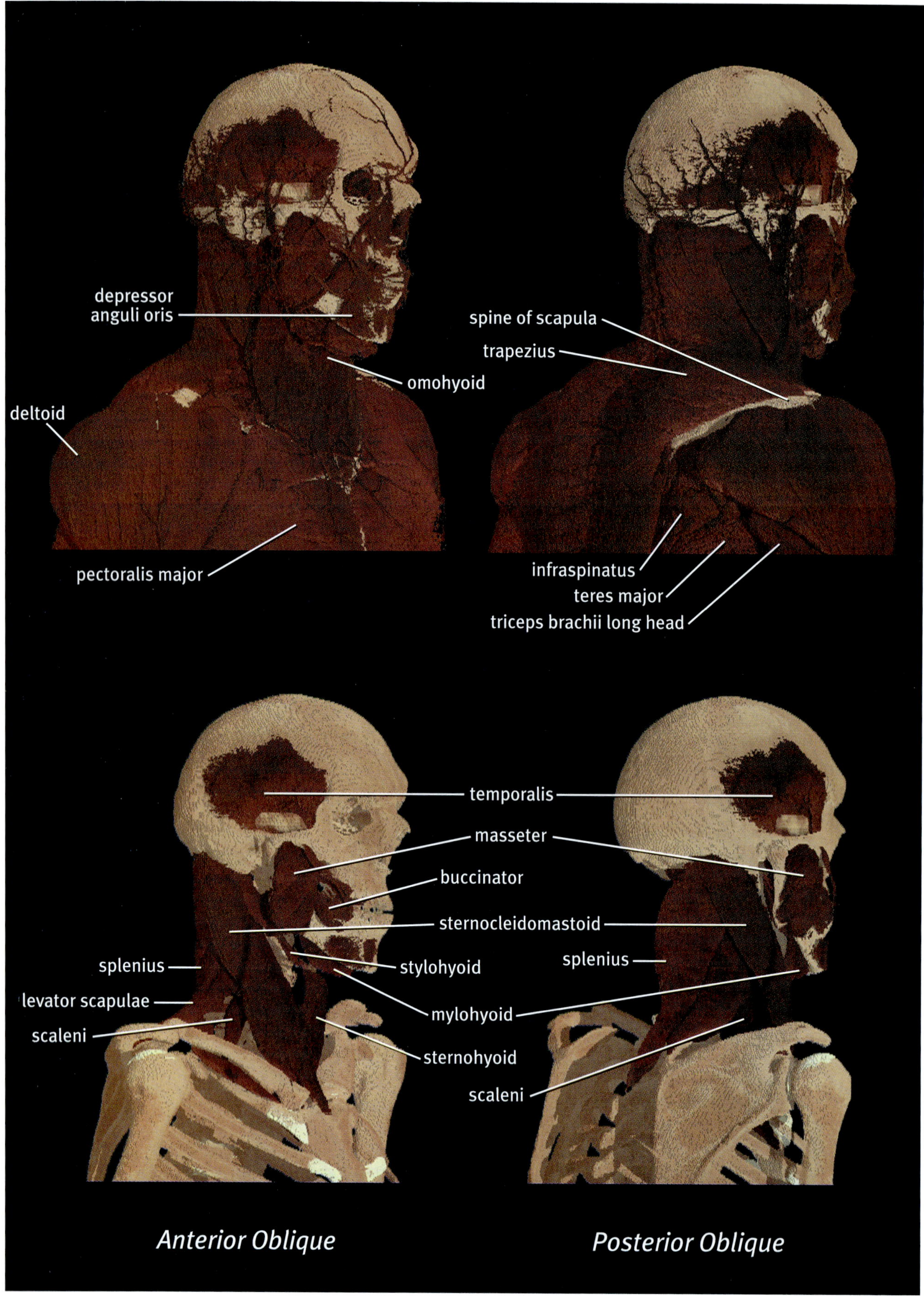

Muscles of the Head and Neck 2

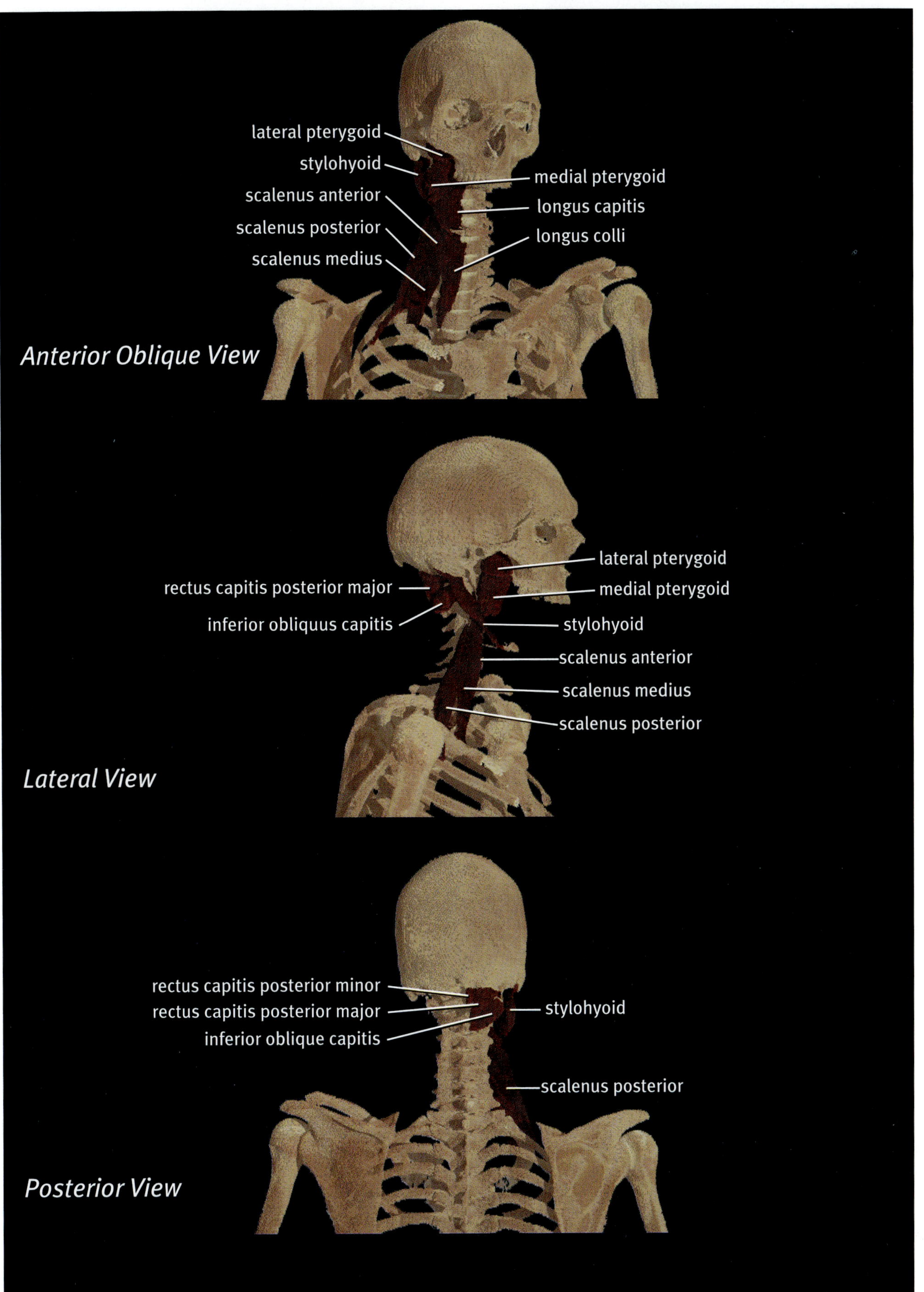

Torso 1

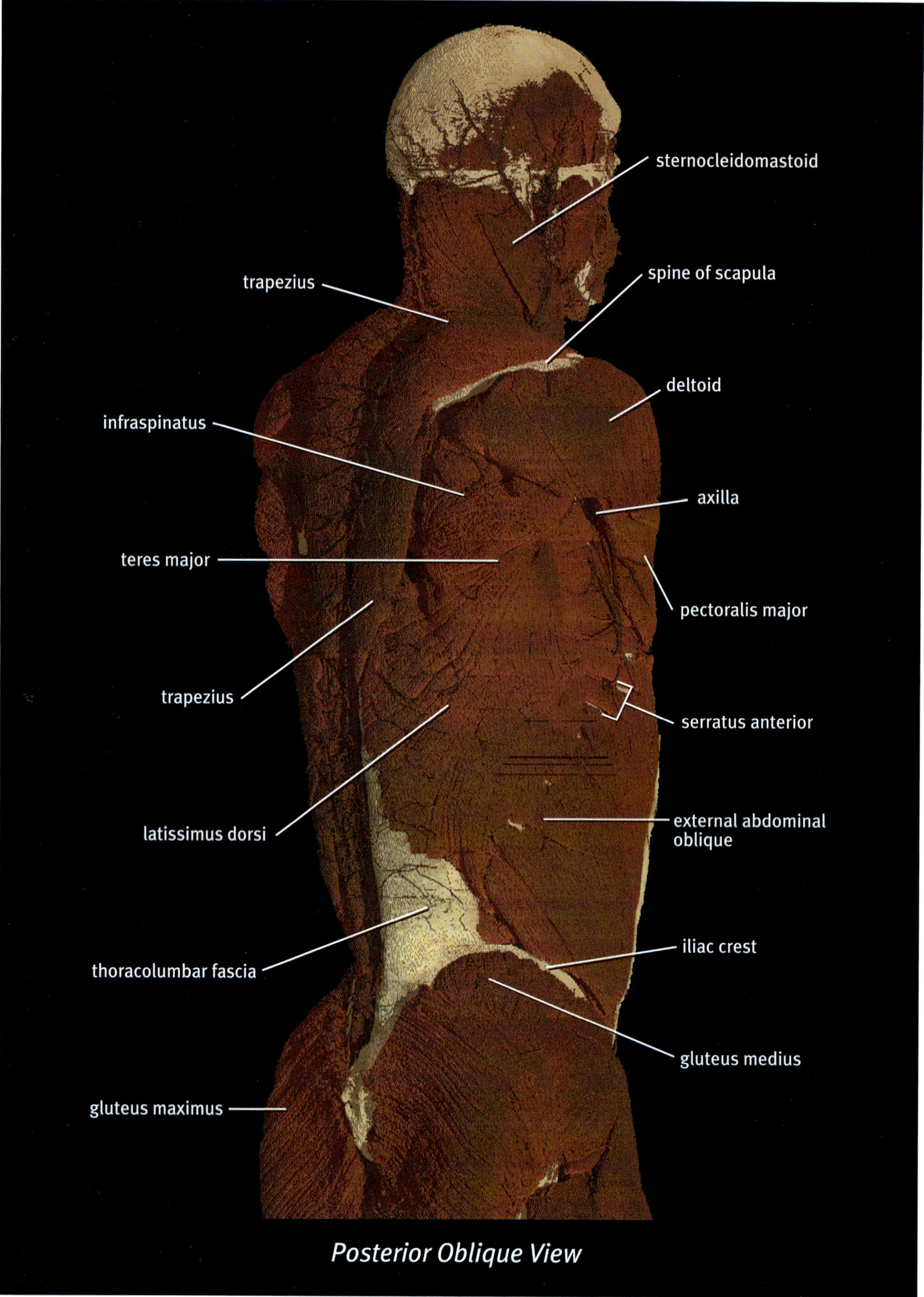

Posterior Oblique View

Torso 2

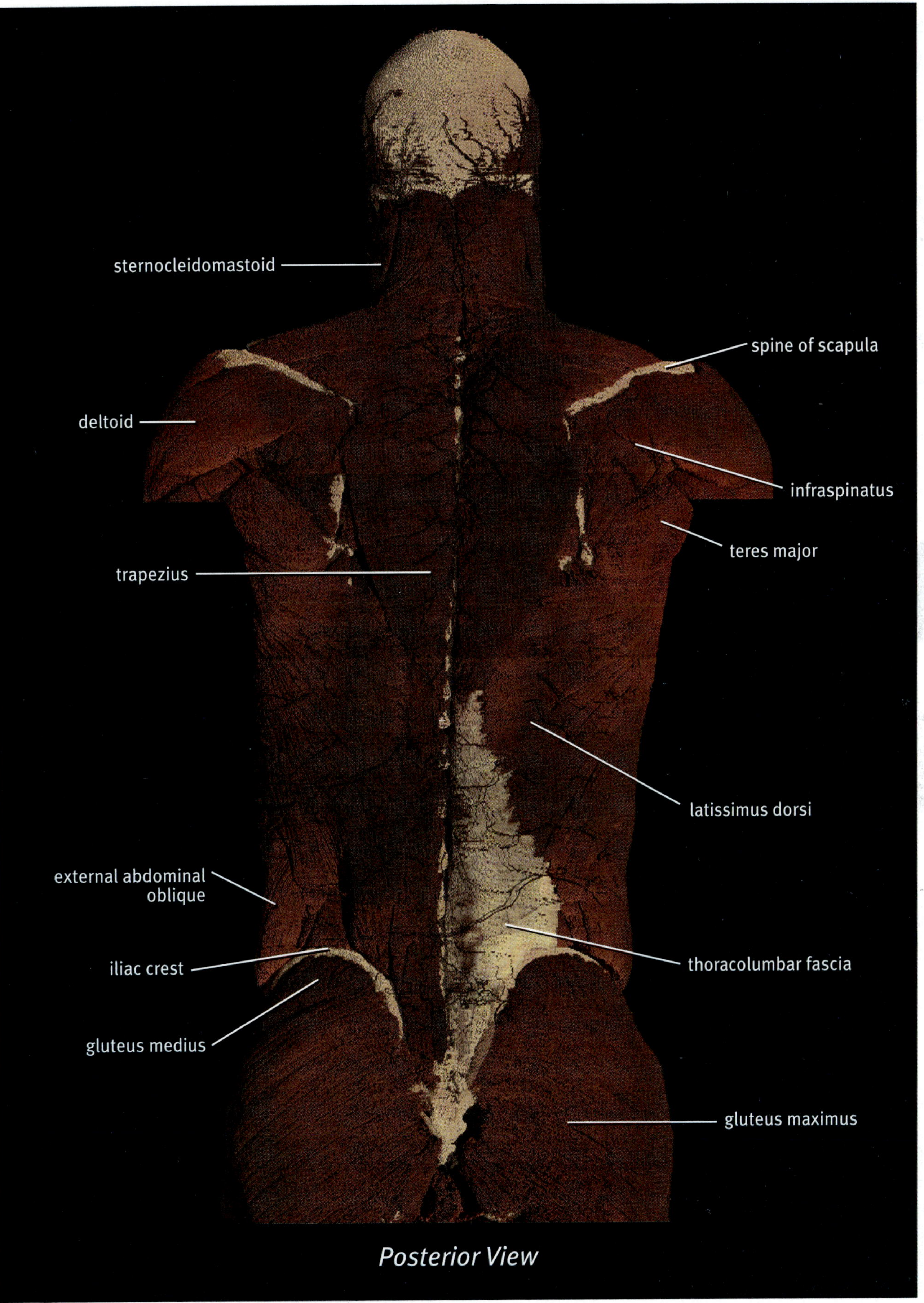

Posterior View

Muscles of the Back

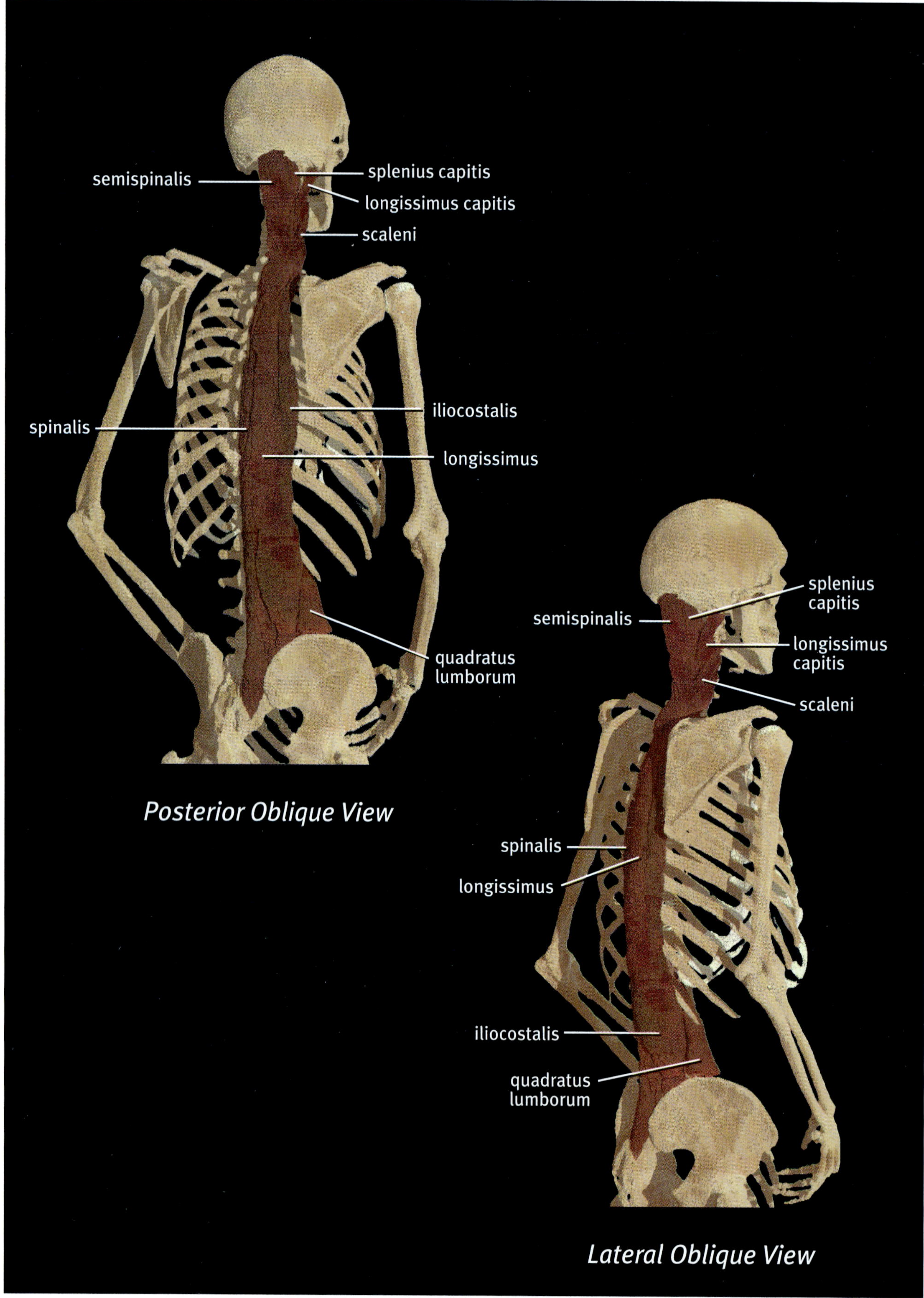

Posterior Oblique View

Lateral Oblique View

Abdominal Muscles 1

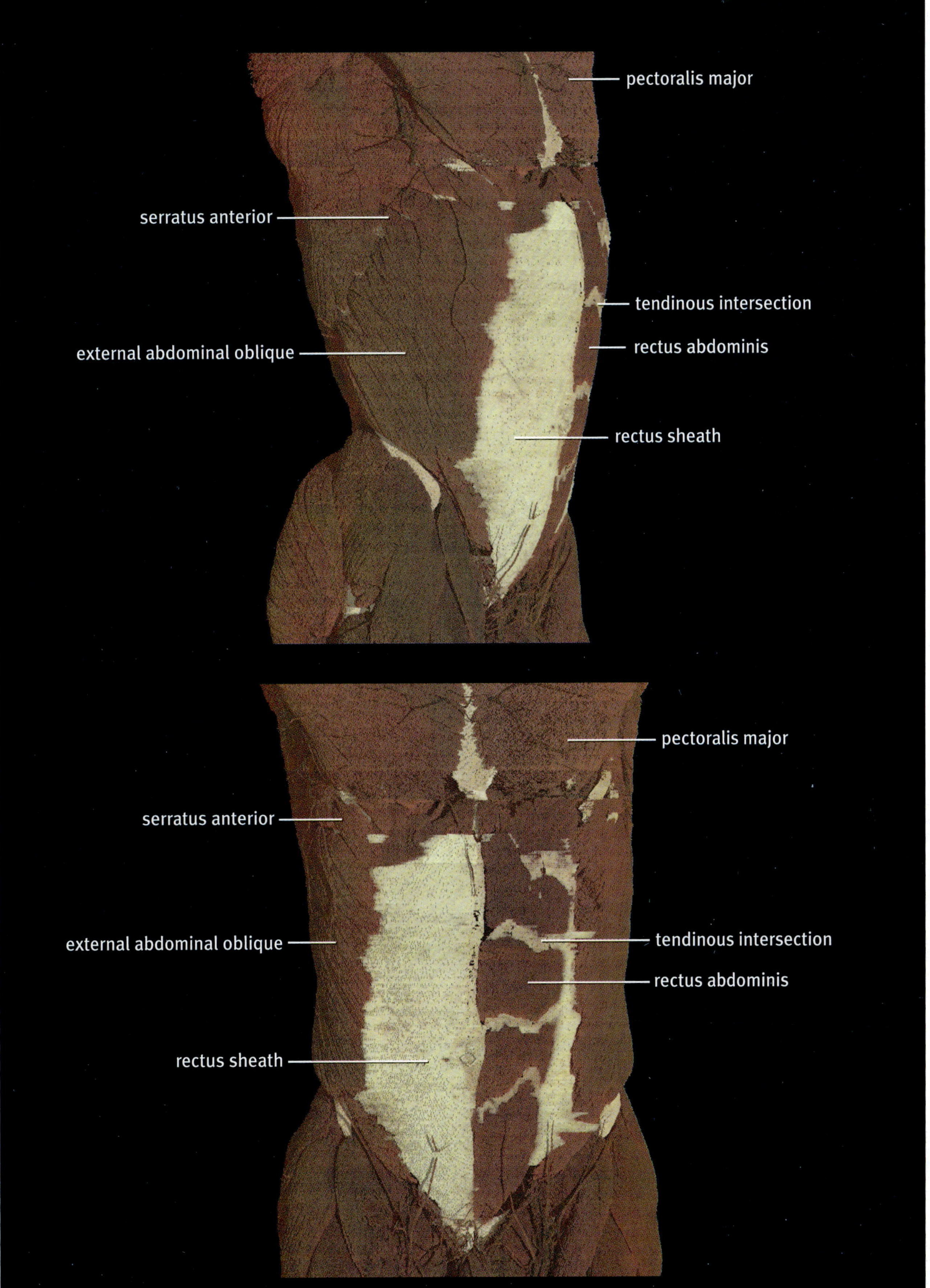

Abdominal Muscles 2

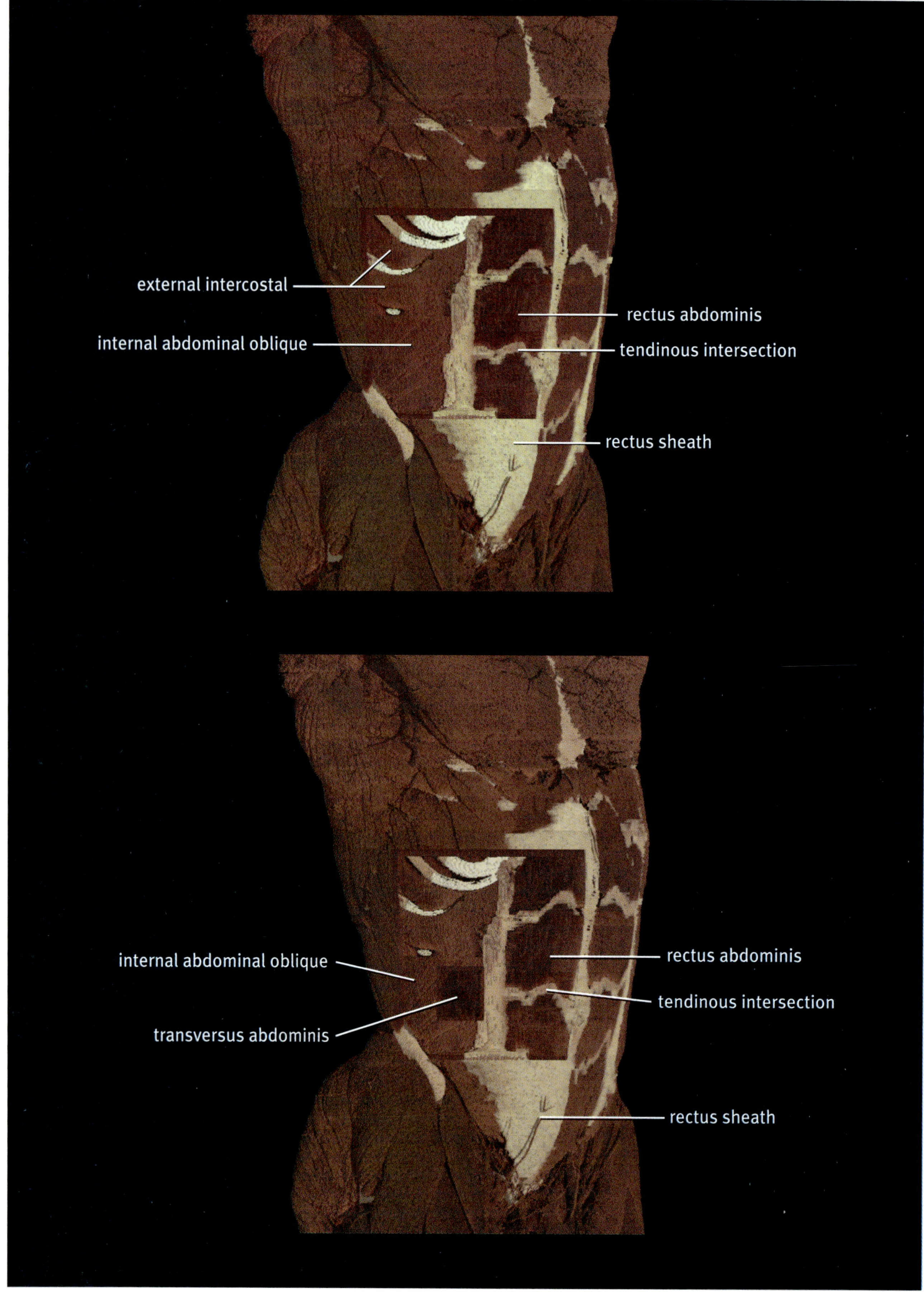

Muscles of the Upper Extremity 1

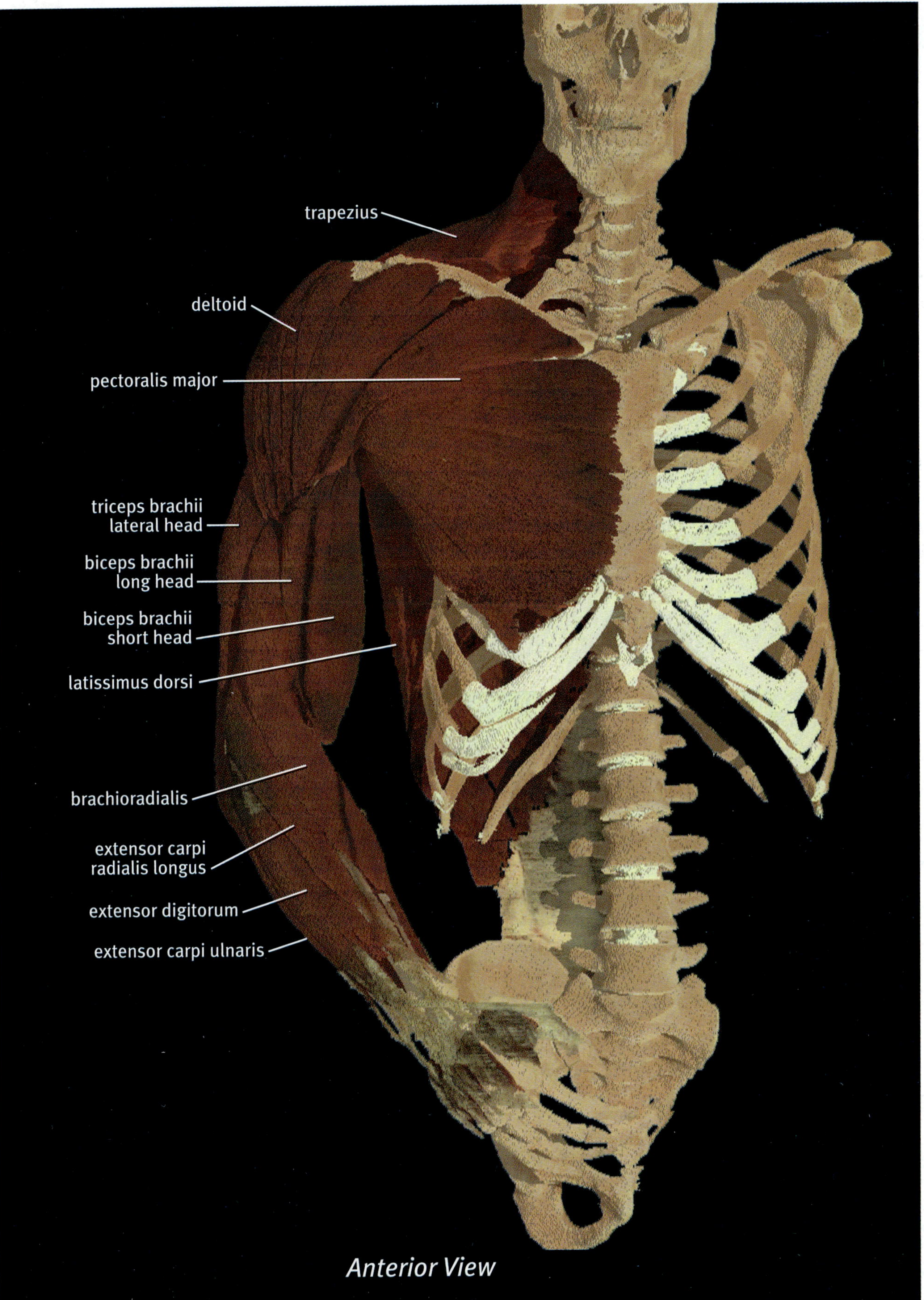

Anterior View

Muscles of the Upper Extremity 2

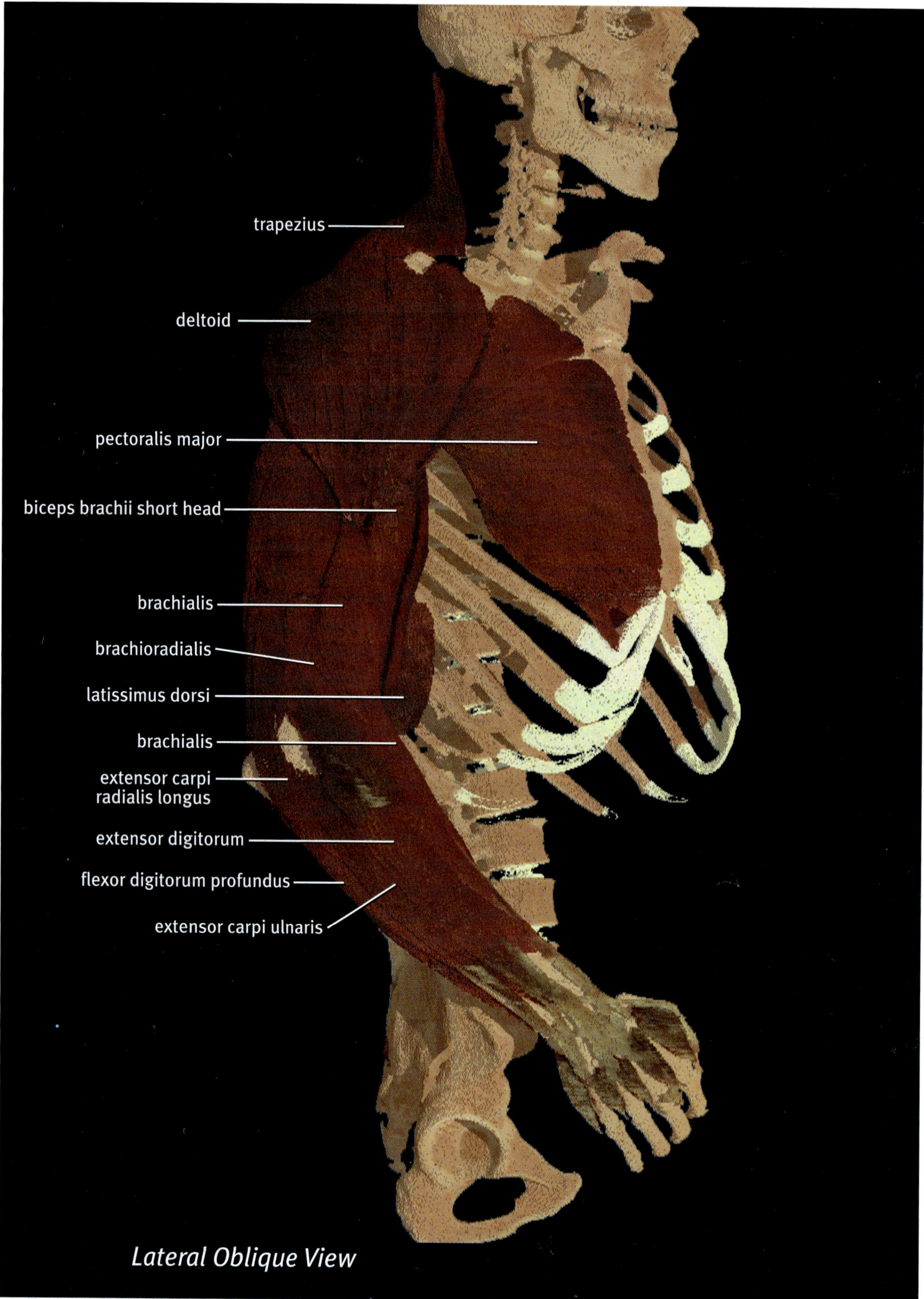

Lateral Oblique View

Muscles of the Upper Extremity 3

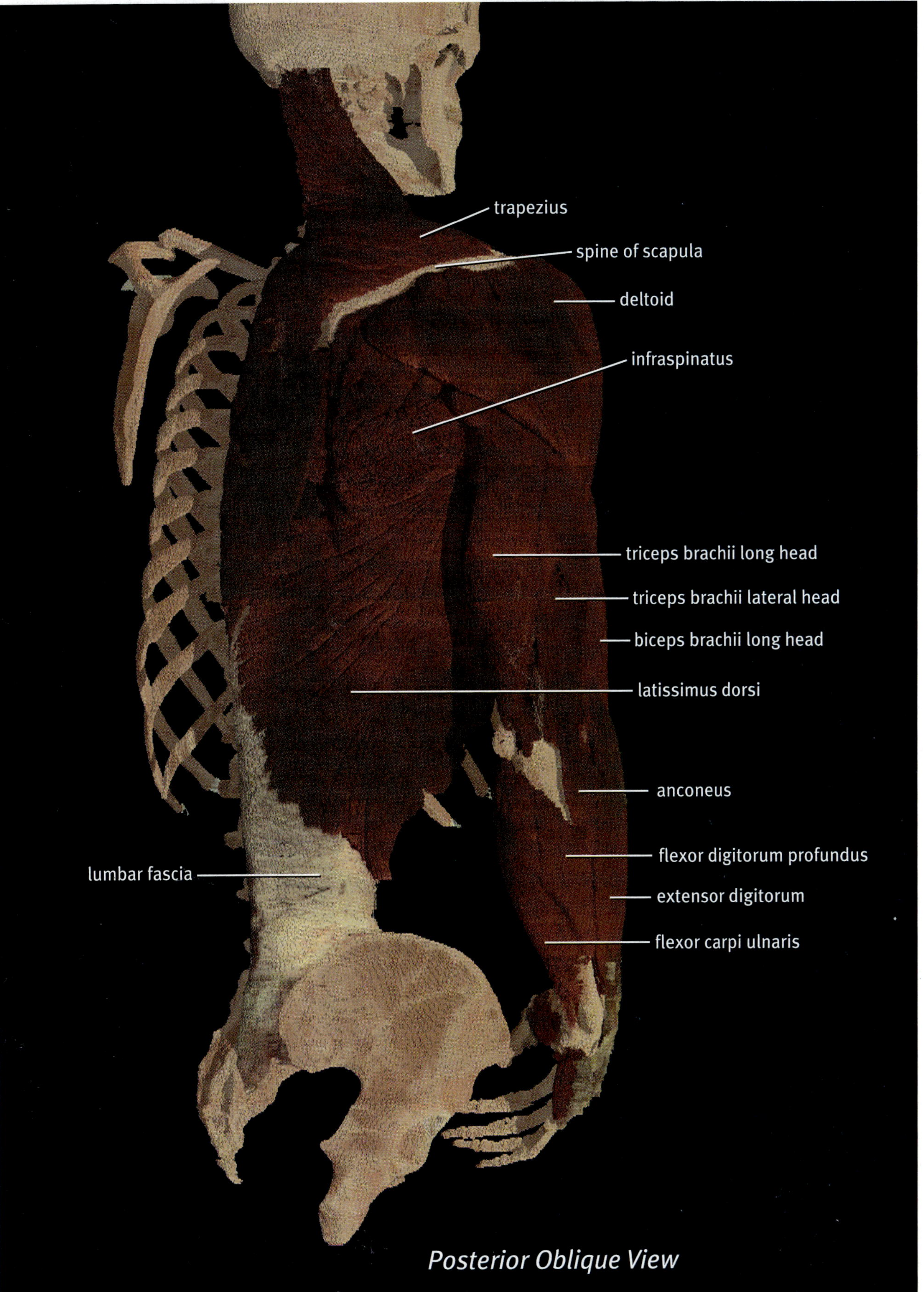

Posterior Oblique View

Cross-sections of the Shoulder

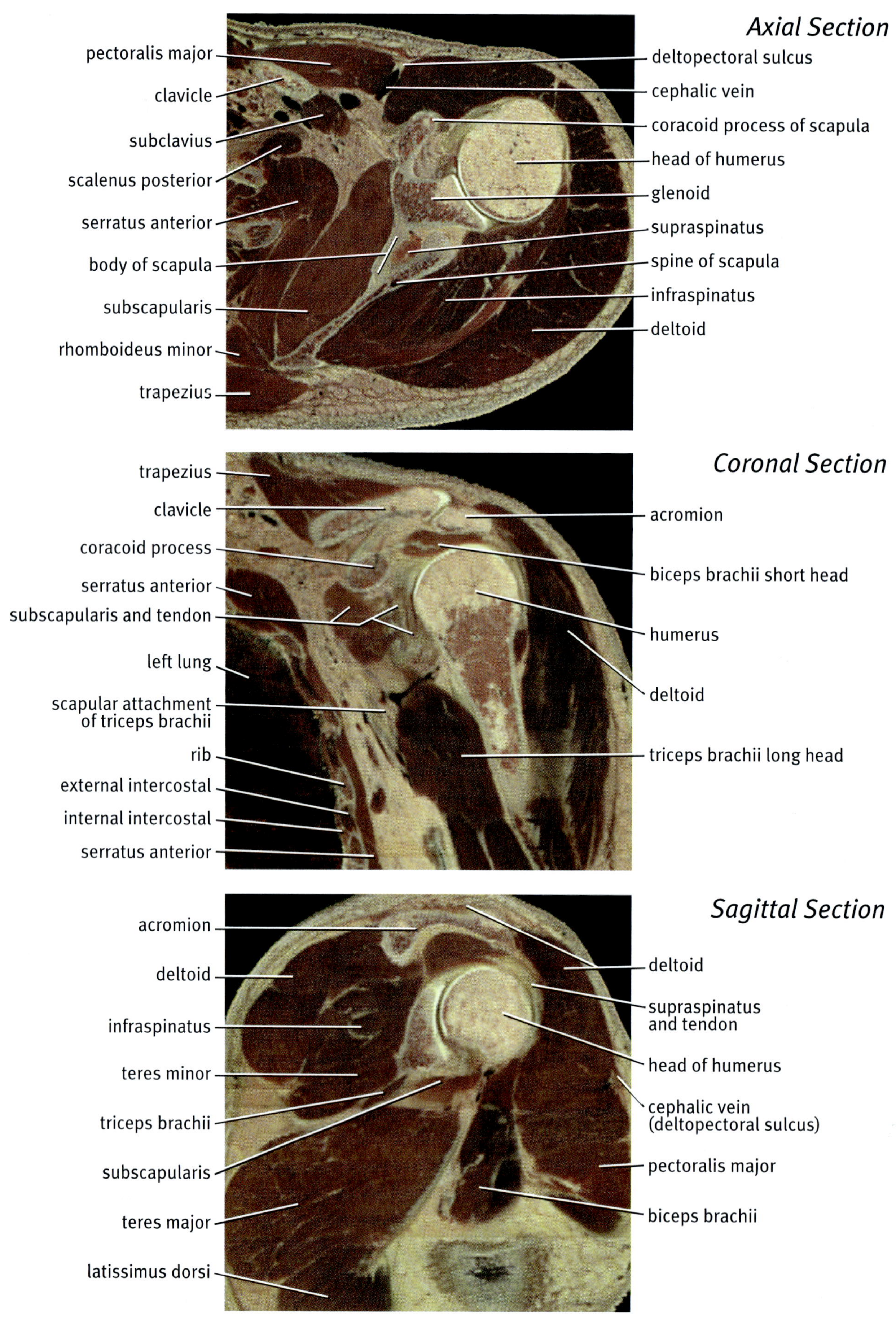

Muscles of the Shoulder and Chest

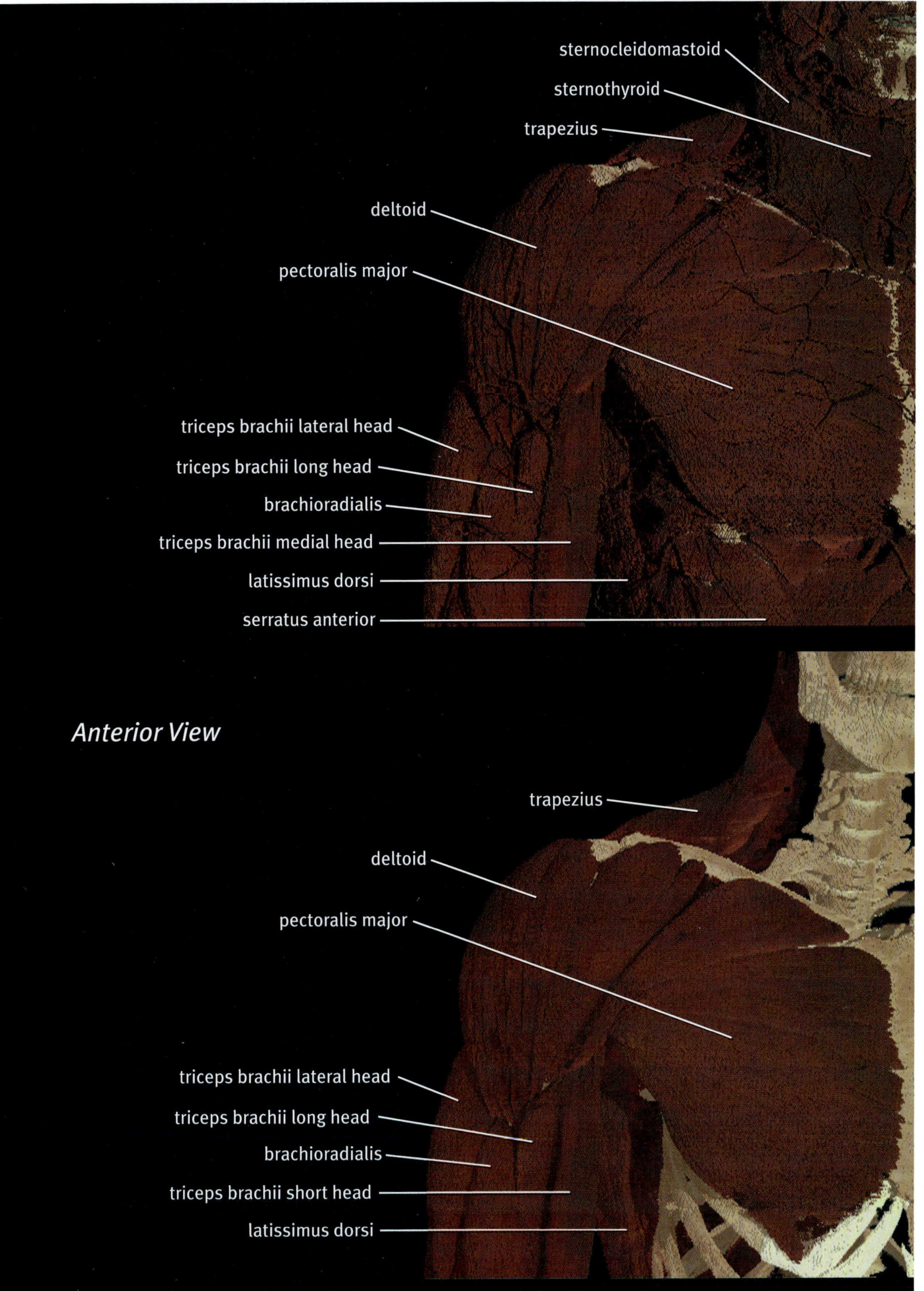

Deep Muscles of the Upper Extremity 1

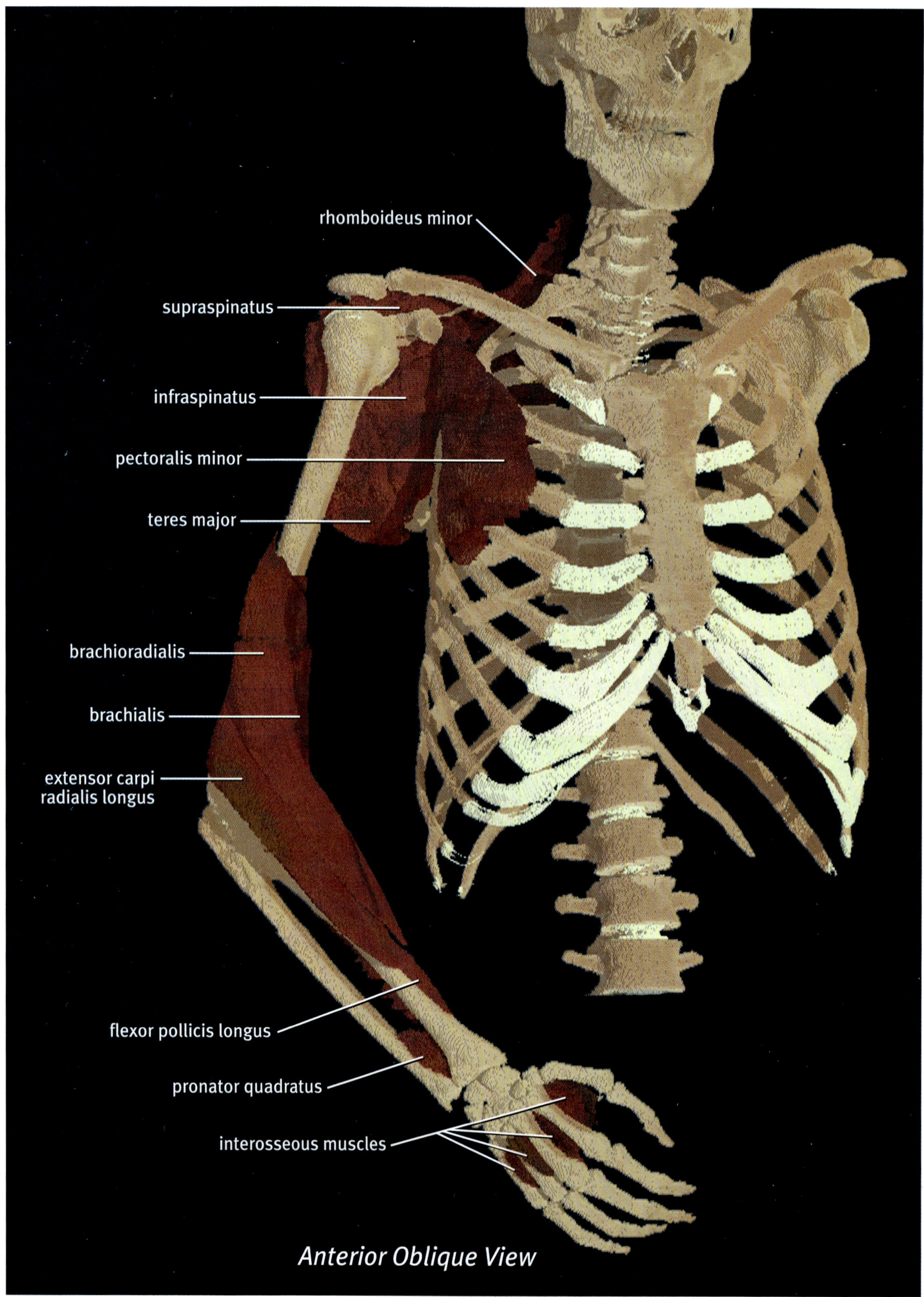

Anterior Oblique View

Deep Muscles of the Upper Extremity 2

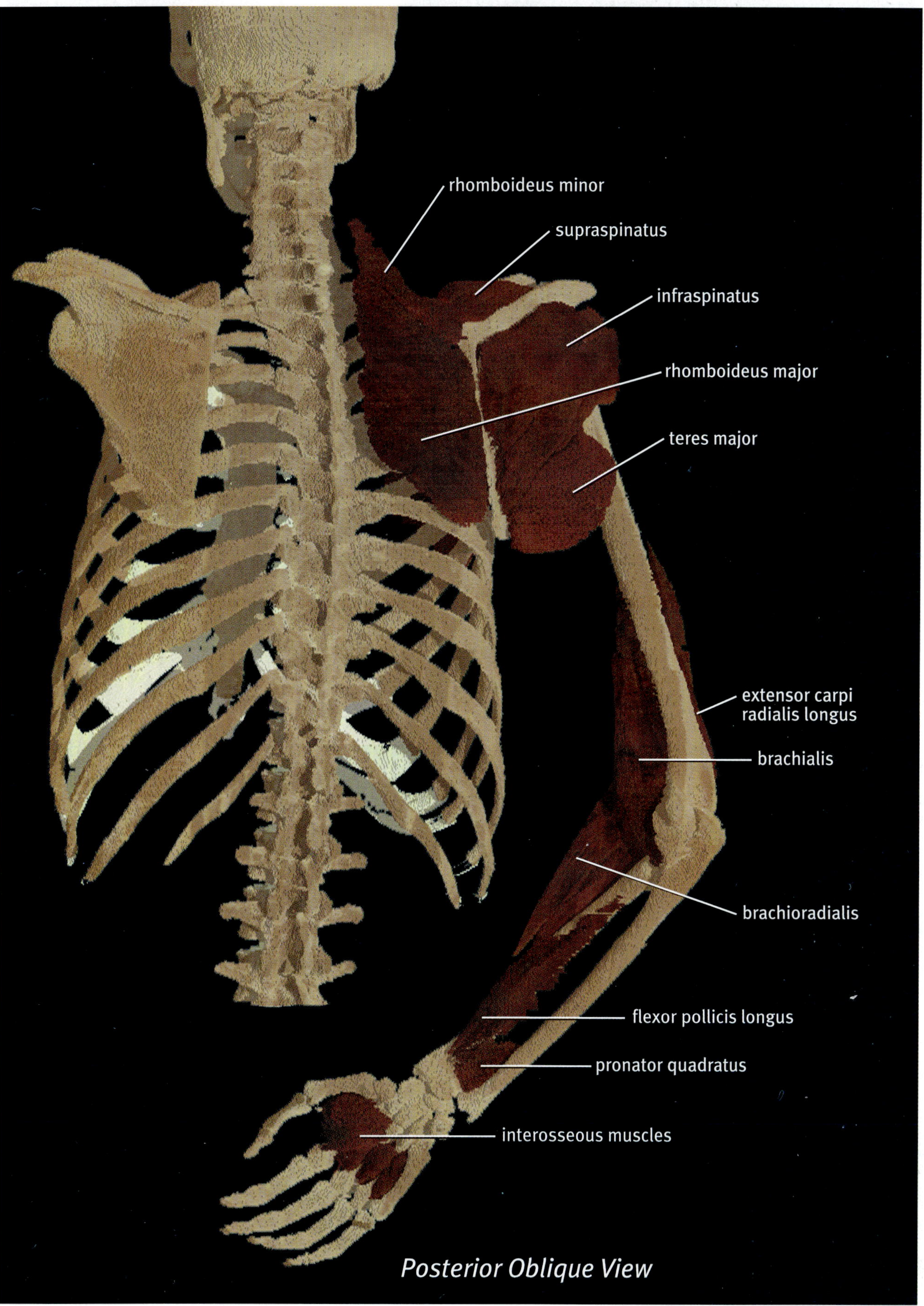

Posterior Oblique View

Muscles of the Forearm

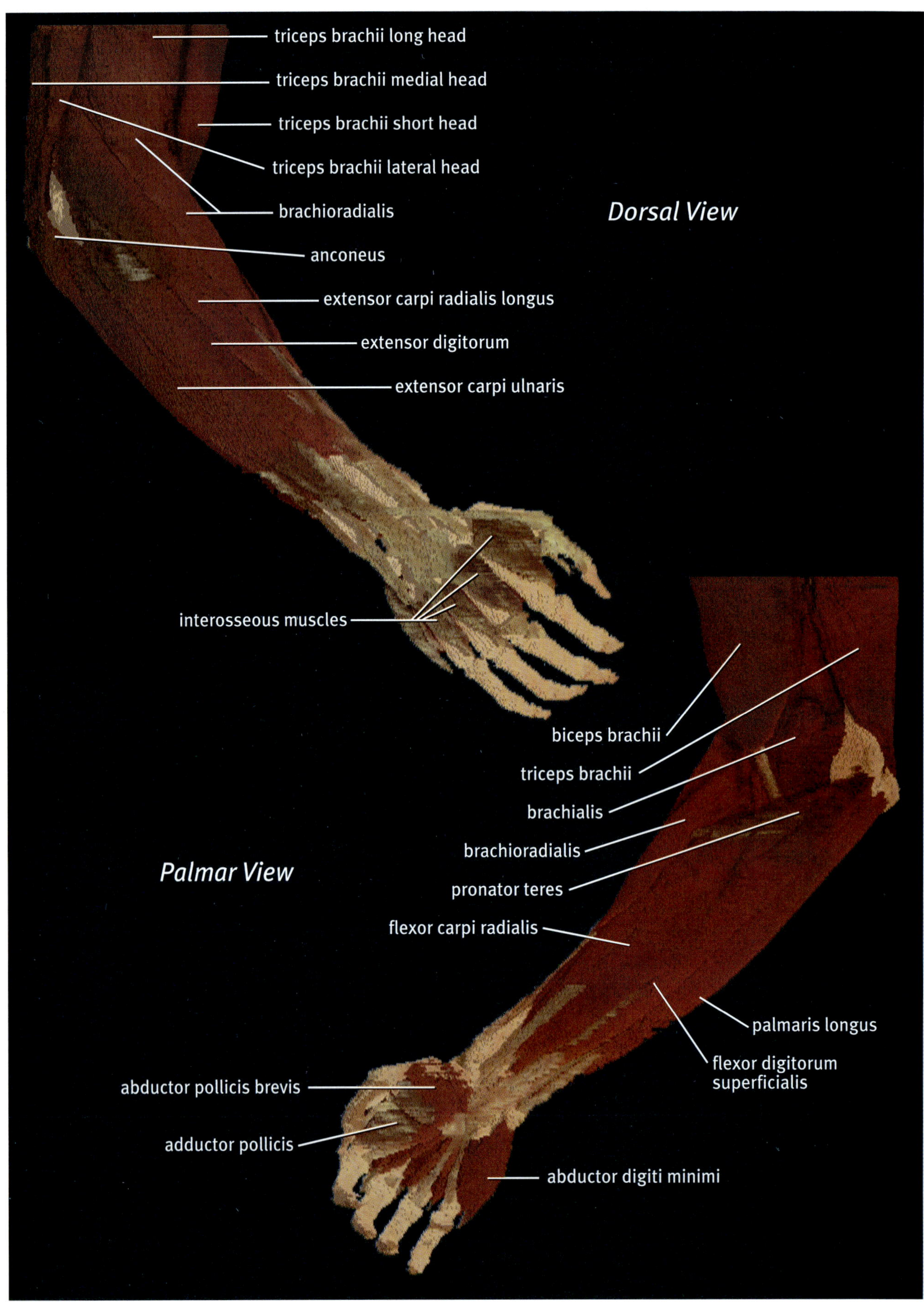

Cross-sections of the Hip

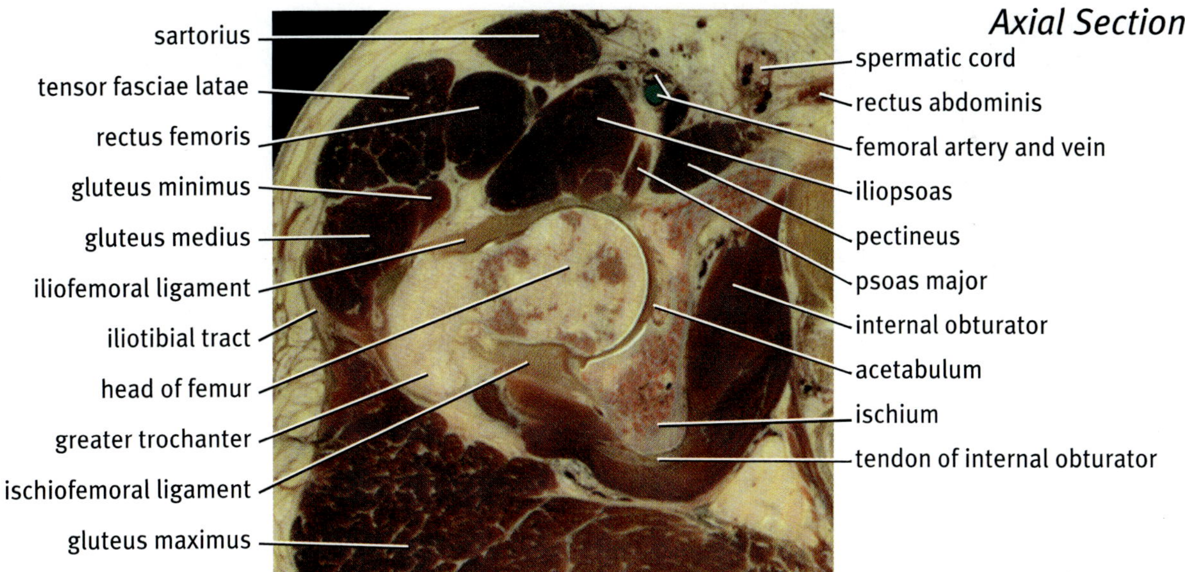

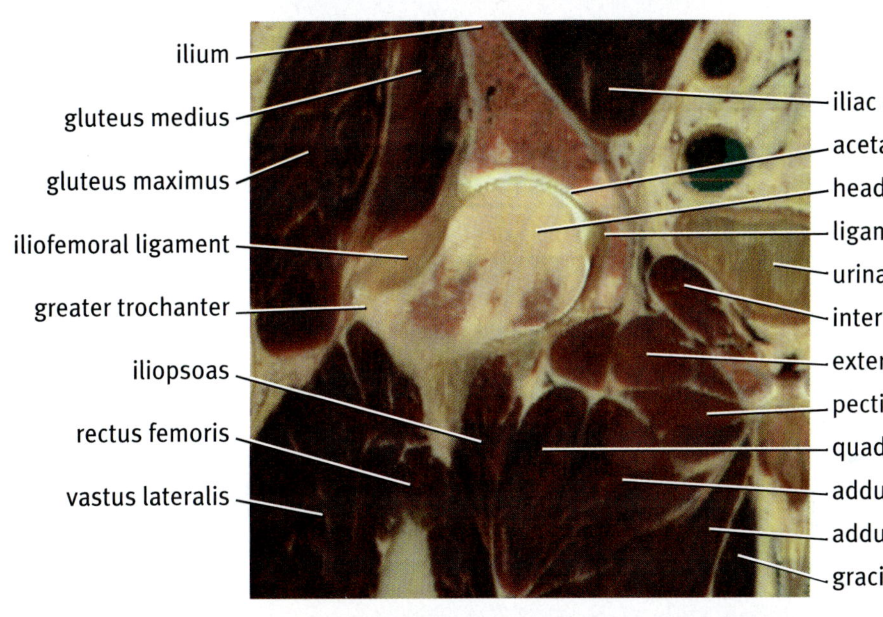

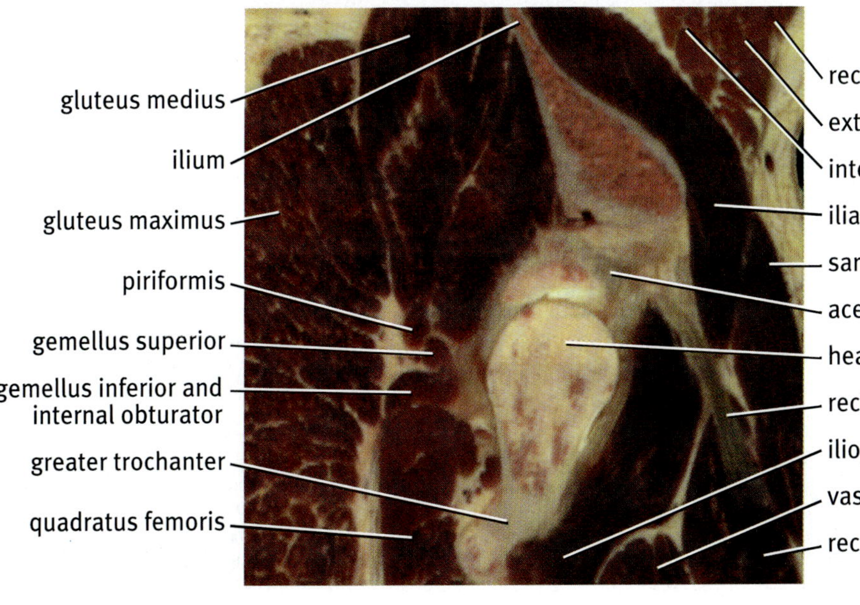

Muscles of the Lower Extremity 1

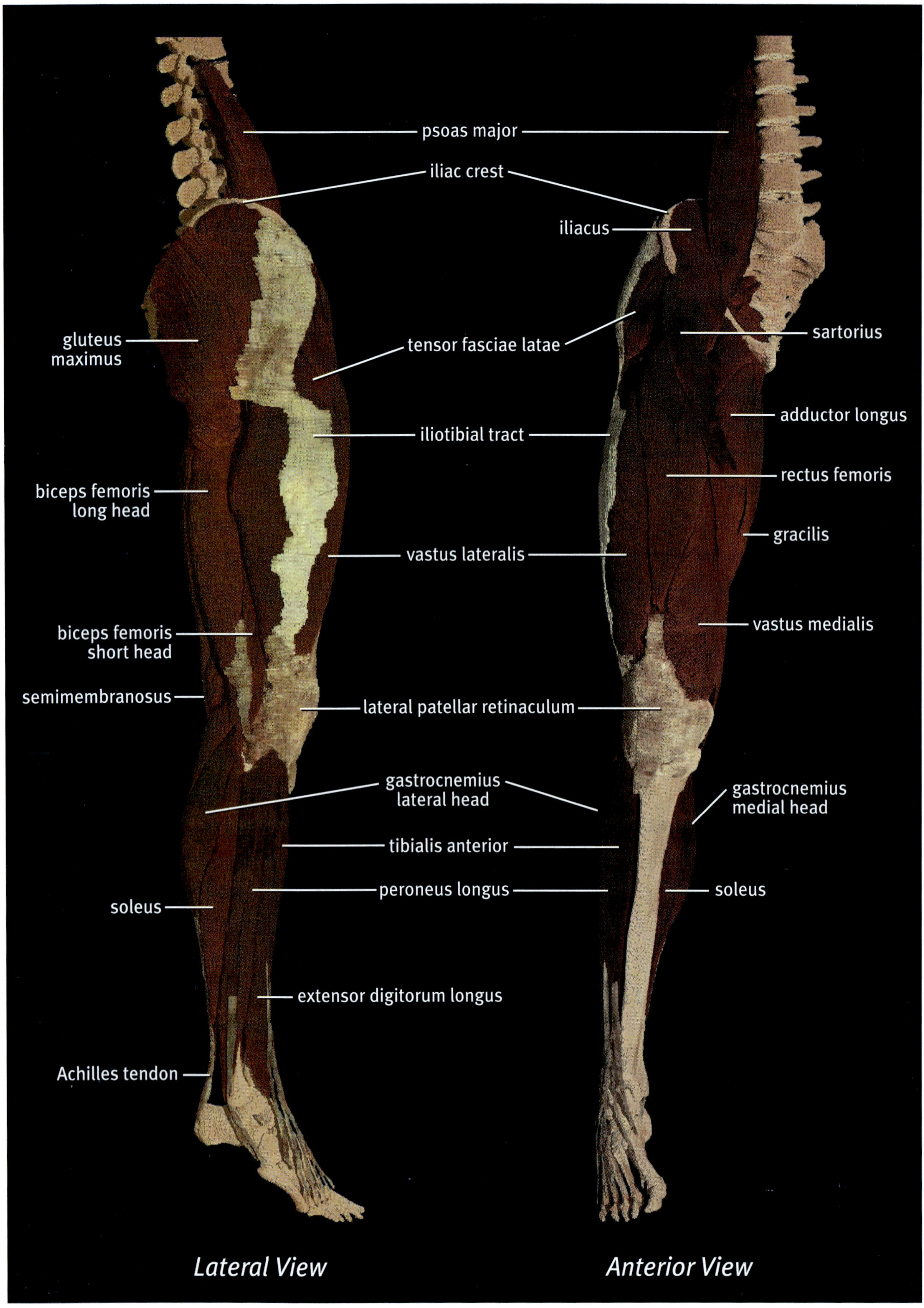

Muscles of the Lower Extremity 2

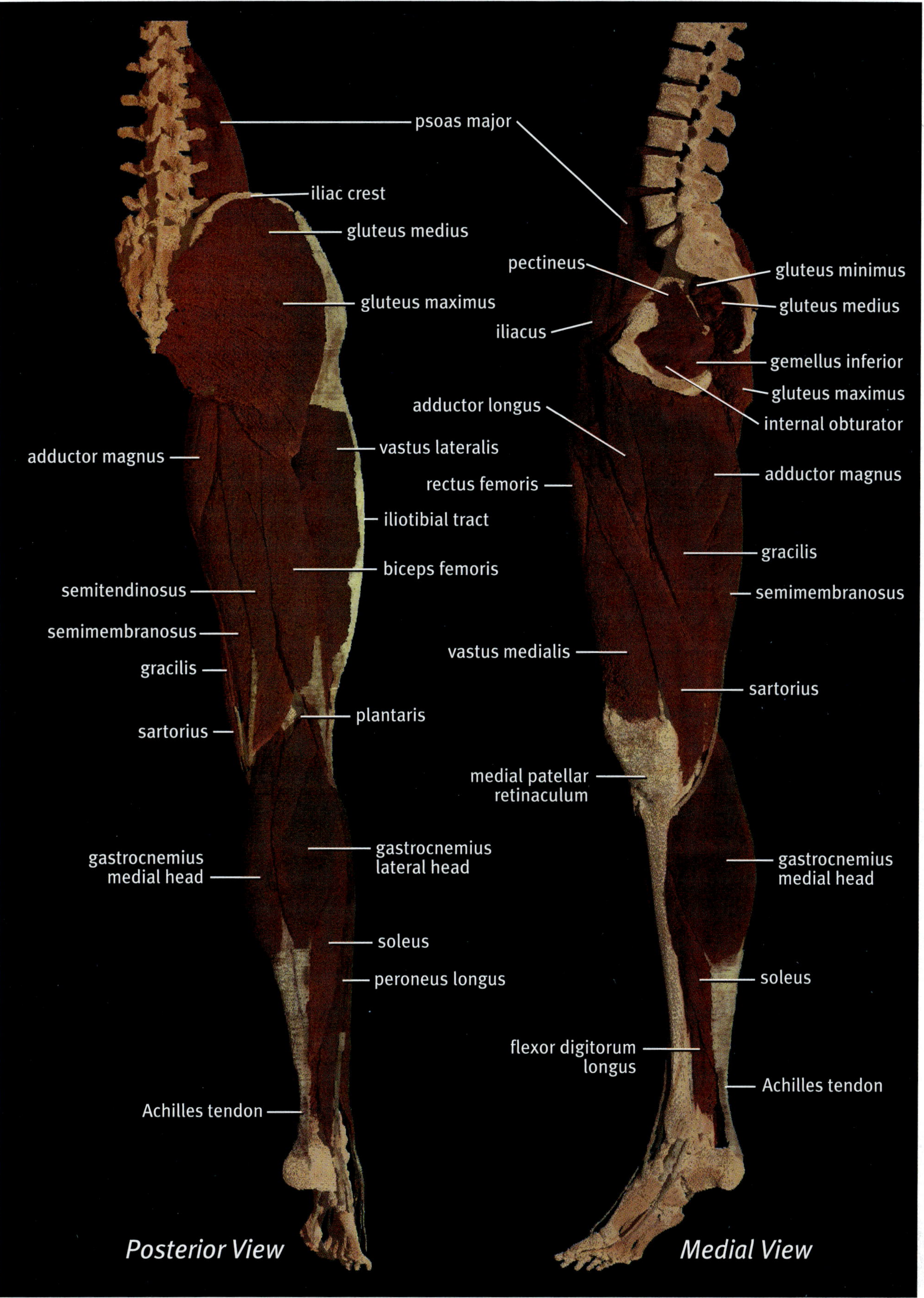

Muscles of the Thigh

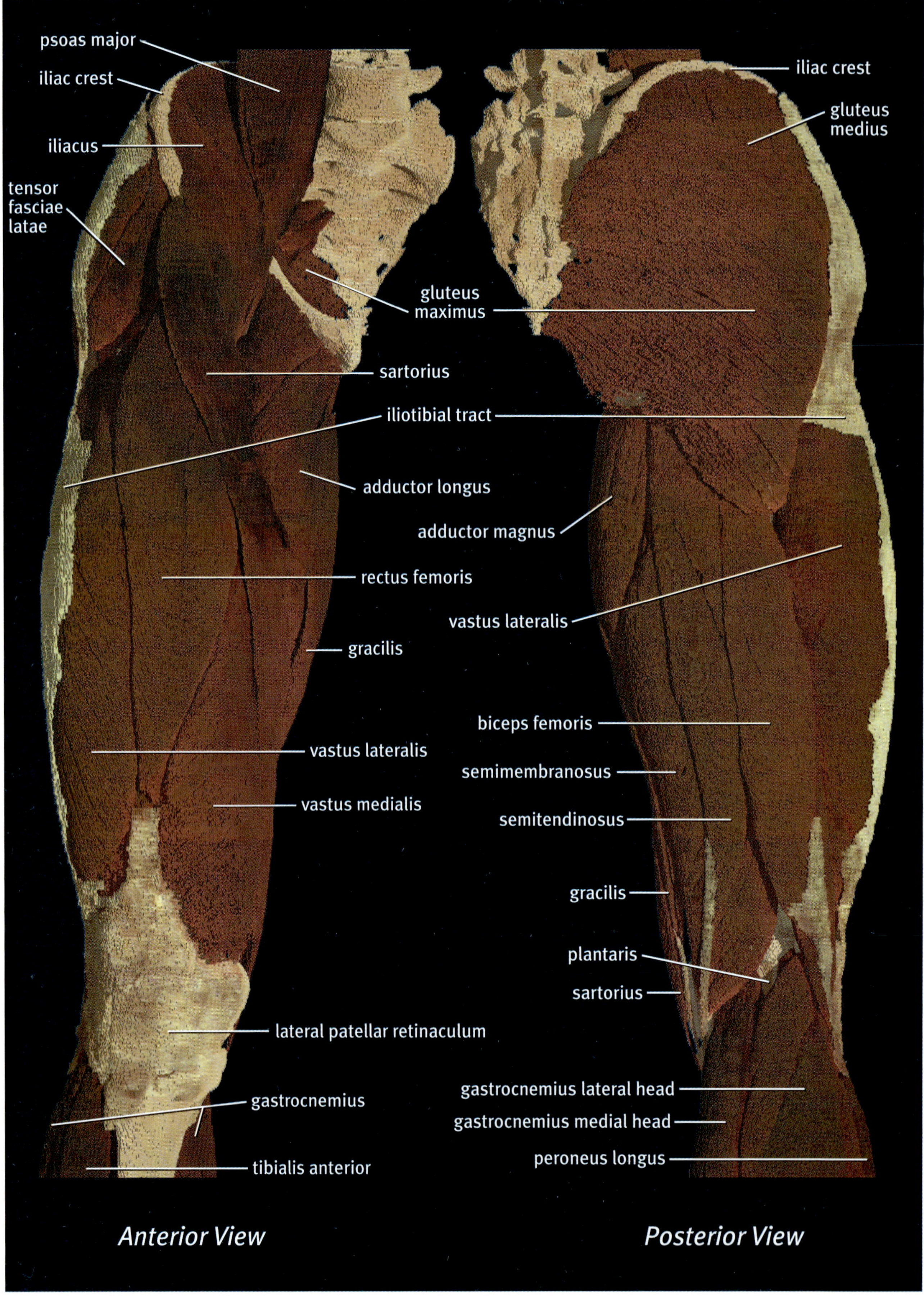

Muscles of the Leg

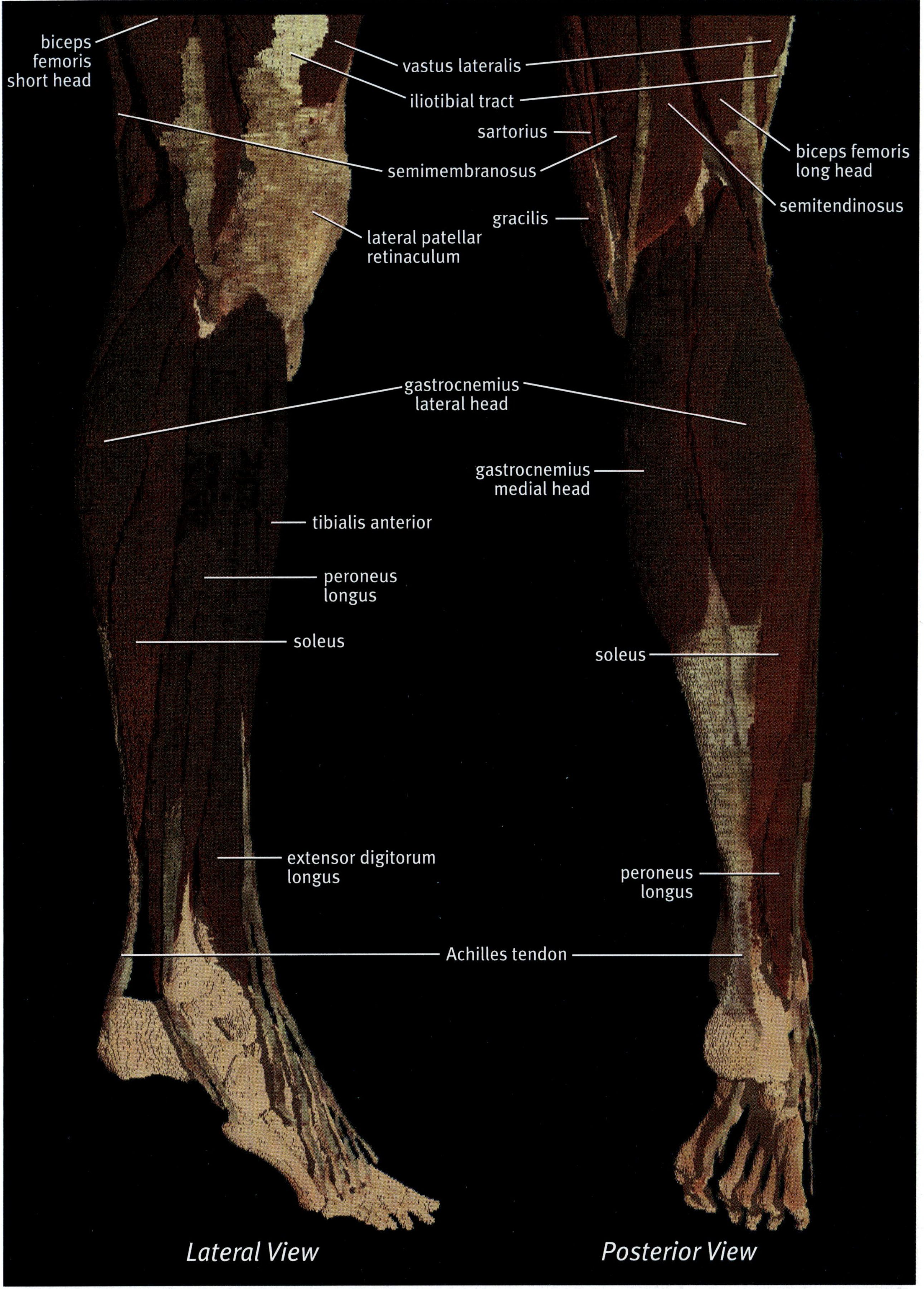

Lateral View *Posterior View*

Deep Muscles of the Lower Extremity

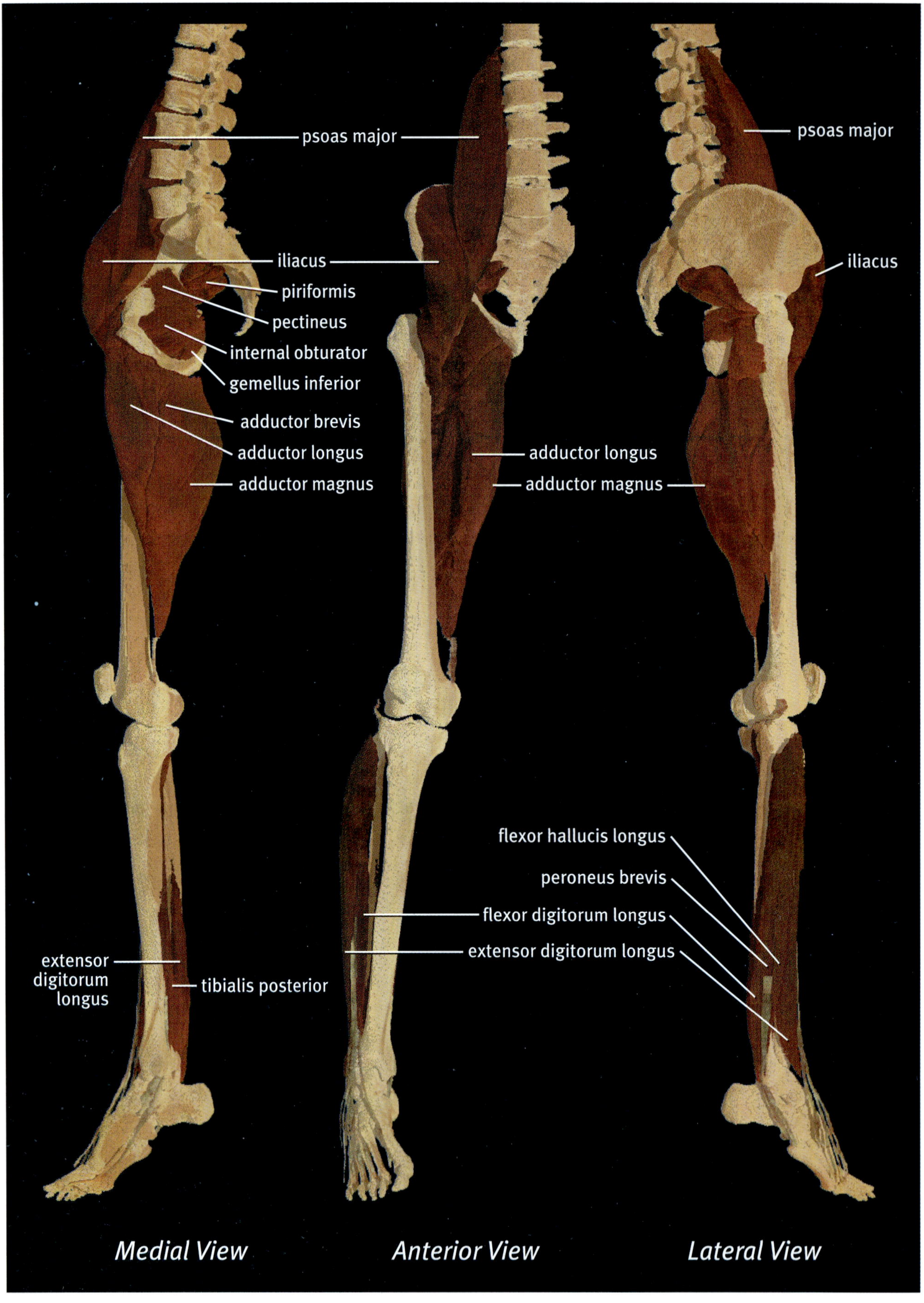

Cross-sections of the Knee

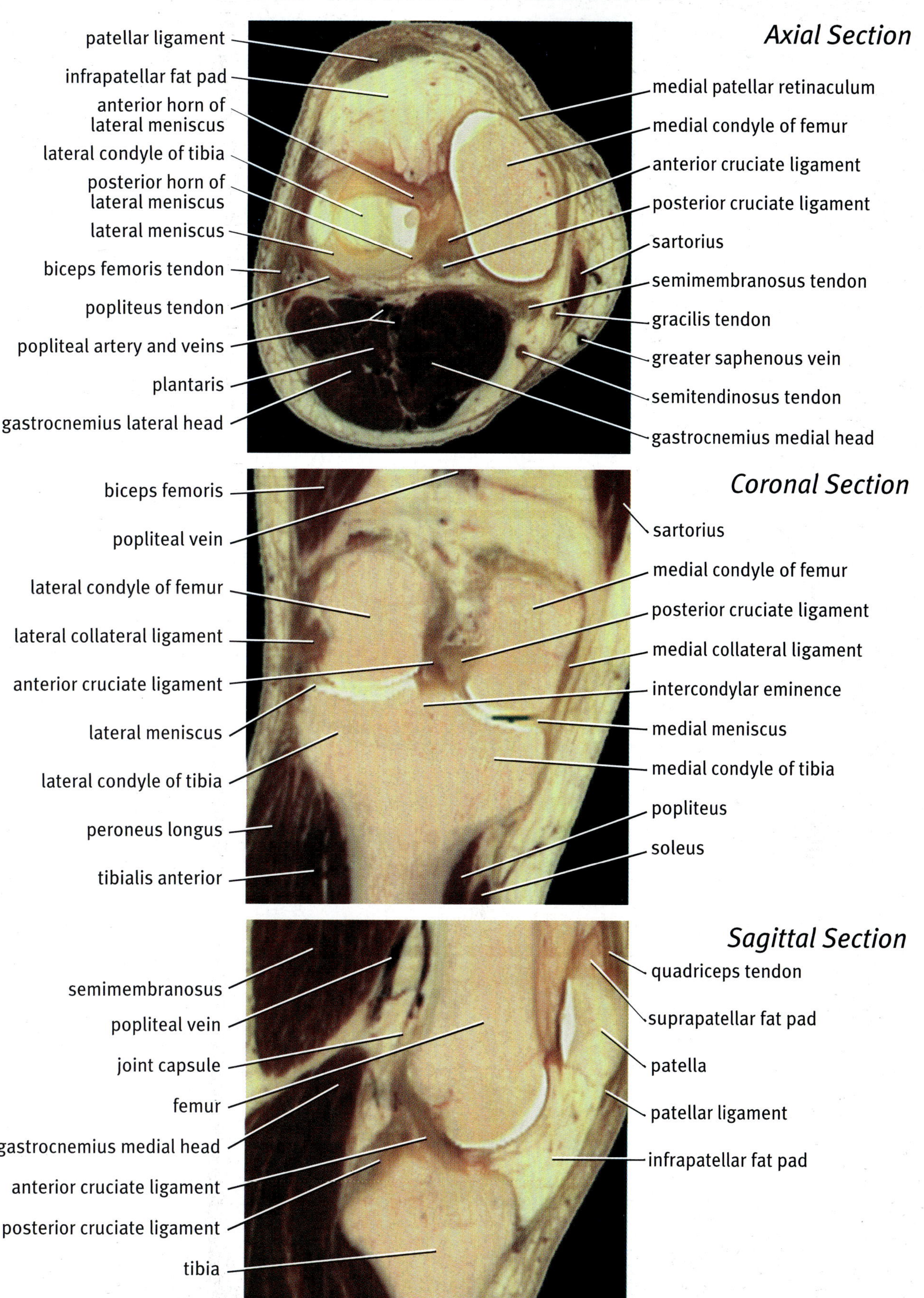

Cross-sections of the Ankle

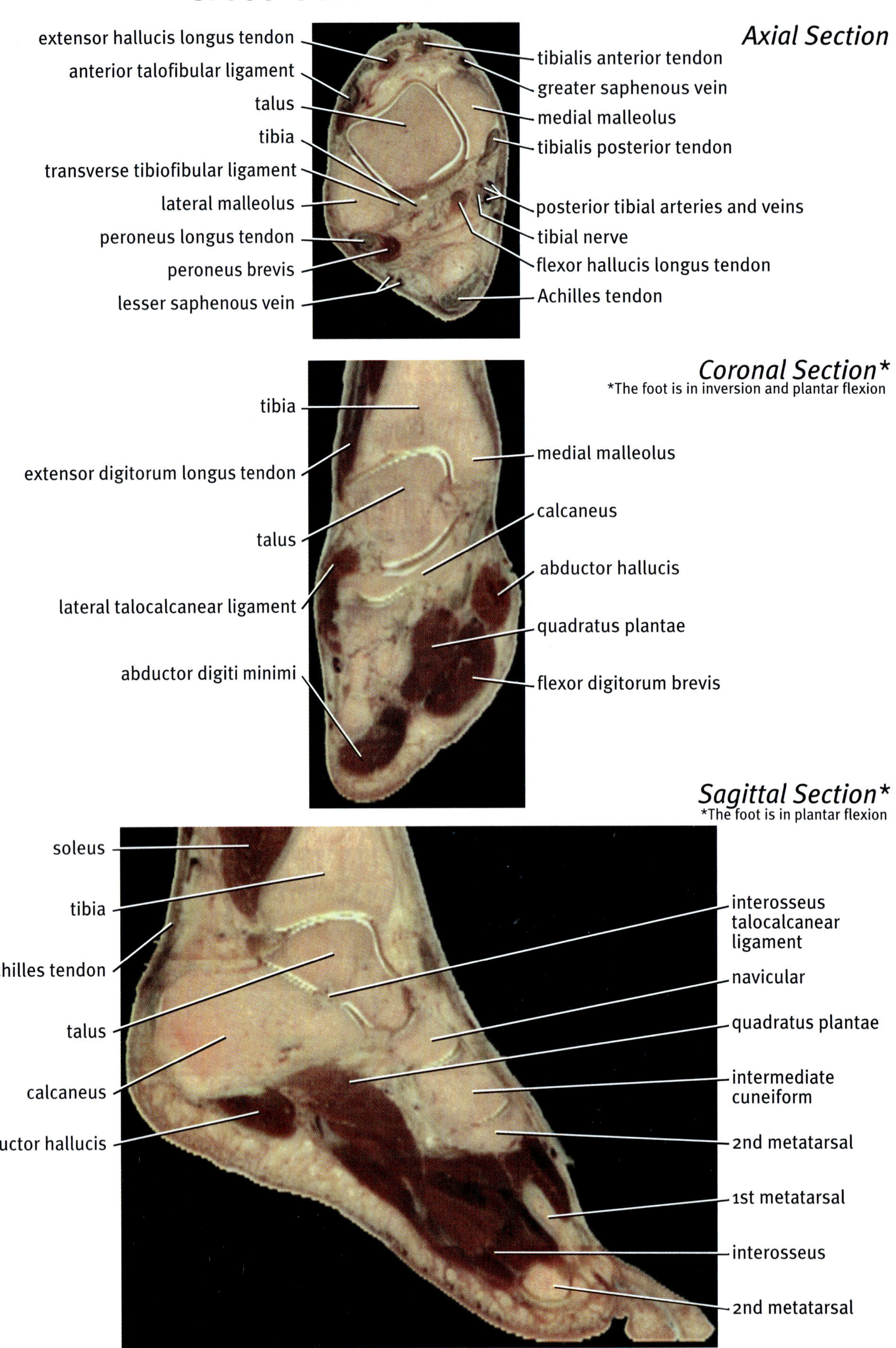

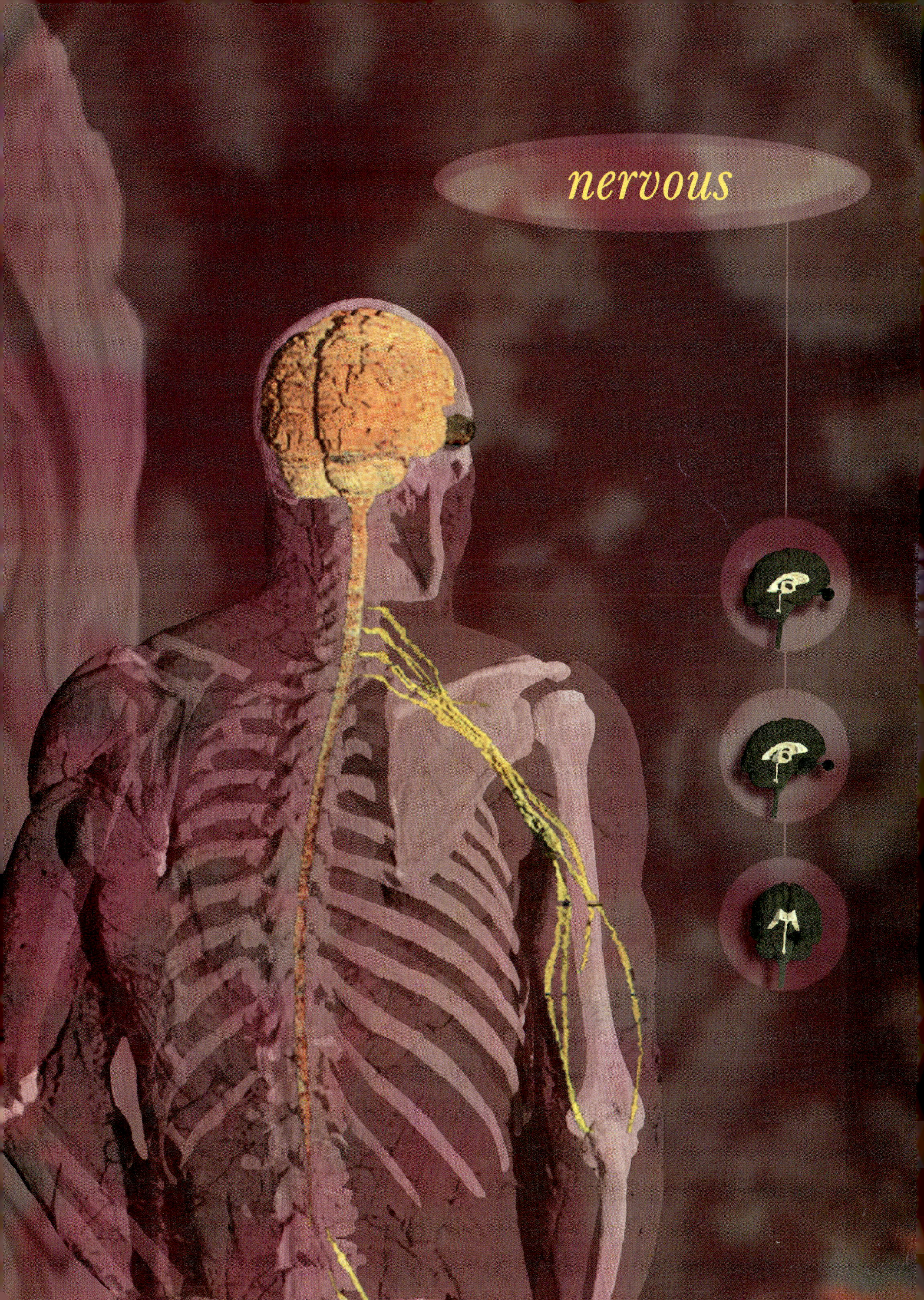
nervous

Nervous System 1

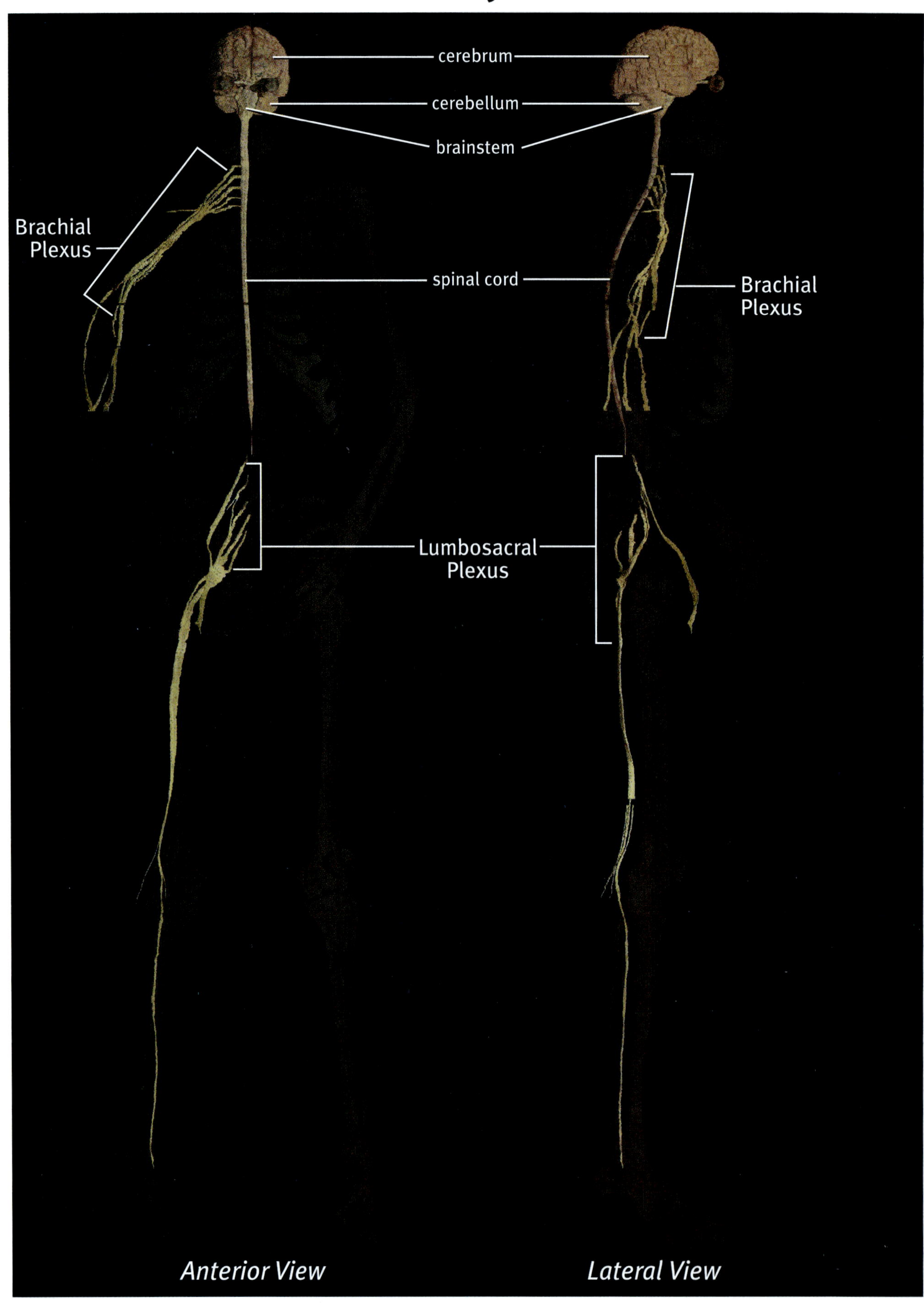

Nervous System 2

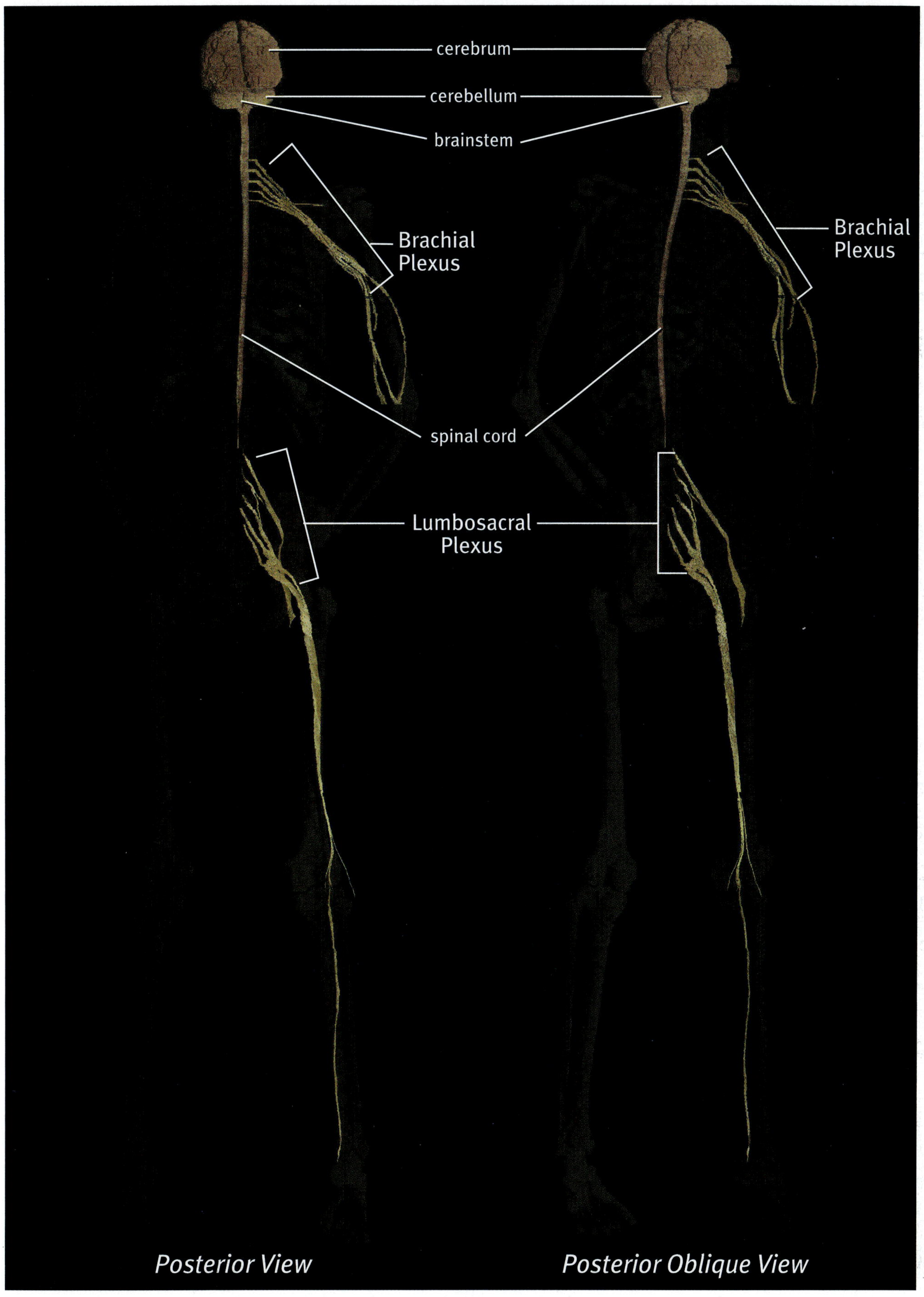

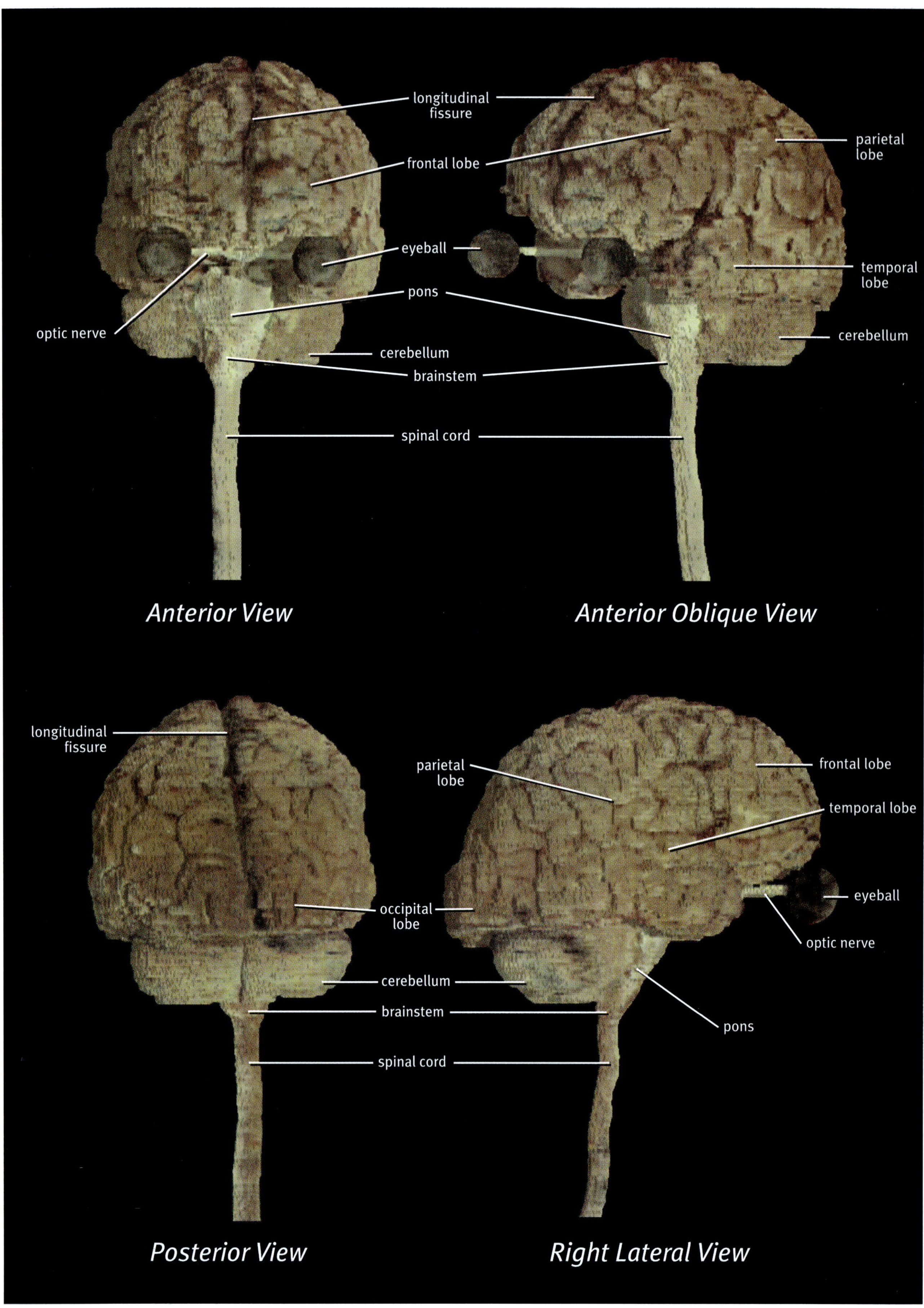
longitudinal fissure
frontal lobe
eyeball
pons
optic nerve
cerebellum
brainstem
spinal cord
parietal lobe
temporal lobe
cerebellum
Anterior View
Anterior Oblique View
longitudinal fissure
parietal lobe
frontal lobe
temporal lobe
eyeball
occipital lobe
optic nerve
cerebellum
brainstem
pons
spinal cord
Posterior View
Right Lateral View

Brain 2

longitudinal fissure
eyeball
temporal pole
optic nerve
optic chiasm
hypophysis (pituitary gland)
pons
brainstem
cerebellum
spinal cord
Inferior Oblique View
optic nerve
diencephalon
cerebrum
cerebellum
left cerebral hemisphere
right cerebral hemisphere
longitudinal fissure
Superior View Full Brain
Superior View
Without Left Hemisphere

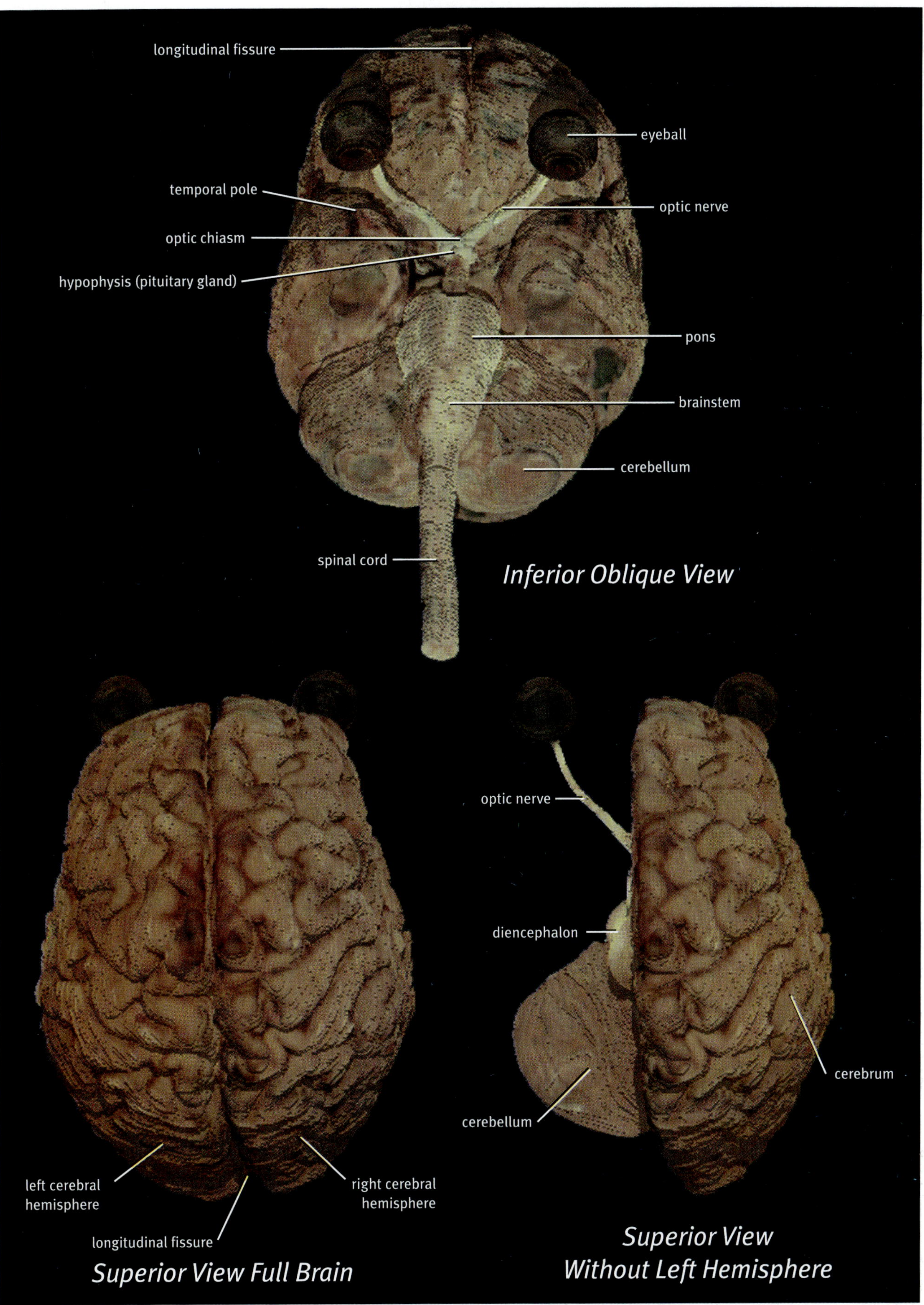

Brain 3

central sulcus
cingulate gyrus
corpus callosum
paraterminal gyrus
eyeball
optic nerve
optic nerve
pons
cerebellum
pons
cerebellum
brainstem
spinal cord
Anterior Oblique View
Left Lateral View
right cerebral hemisphere
corpus callosum
optic nerve
eyeball
cerebellum
brainstem
spinal cord
Posterior View
Posterior Oblique View

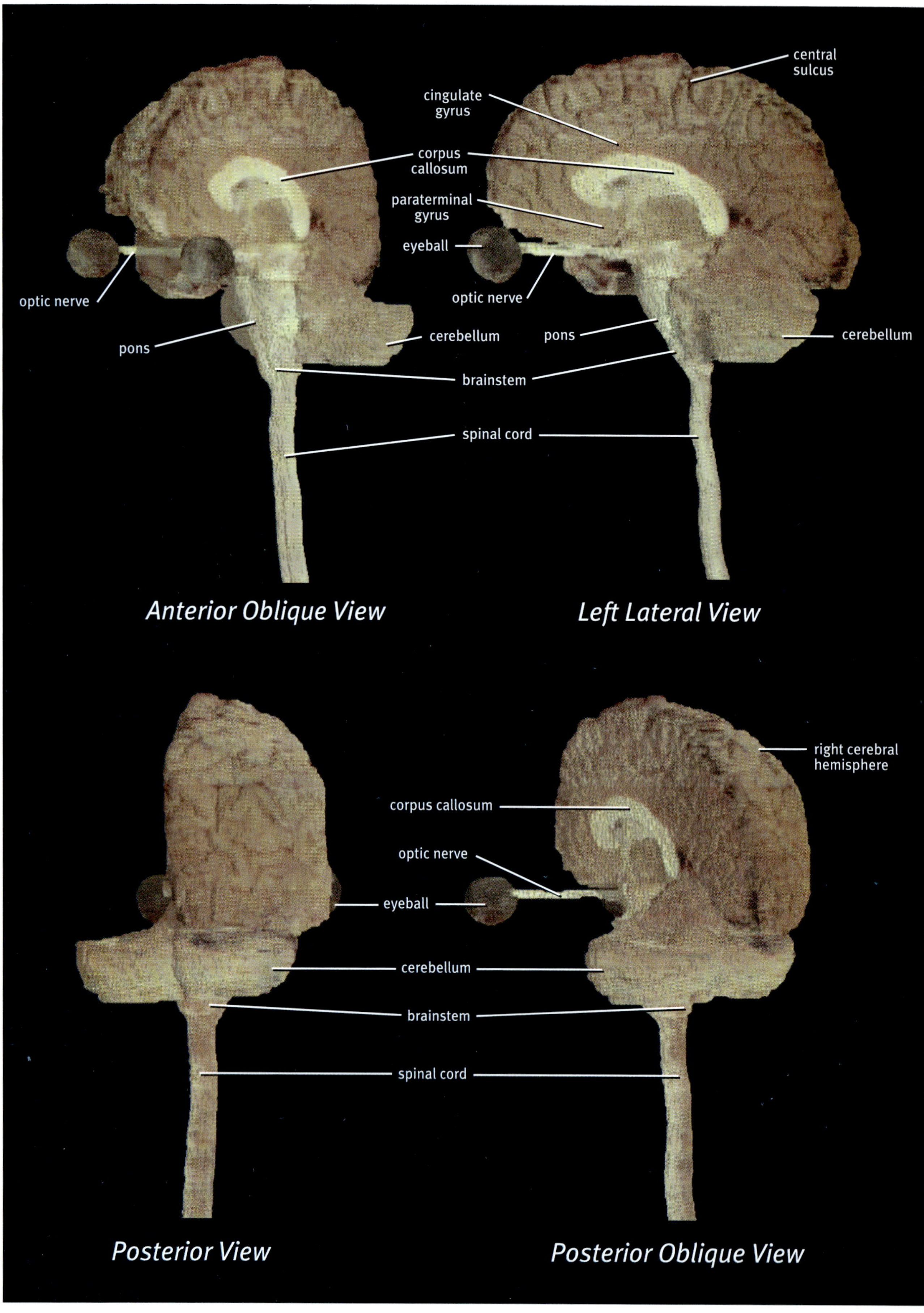

Ventricles of the Brain

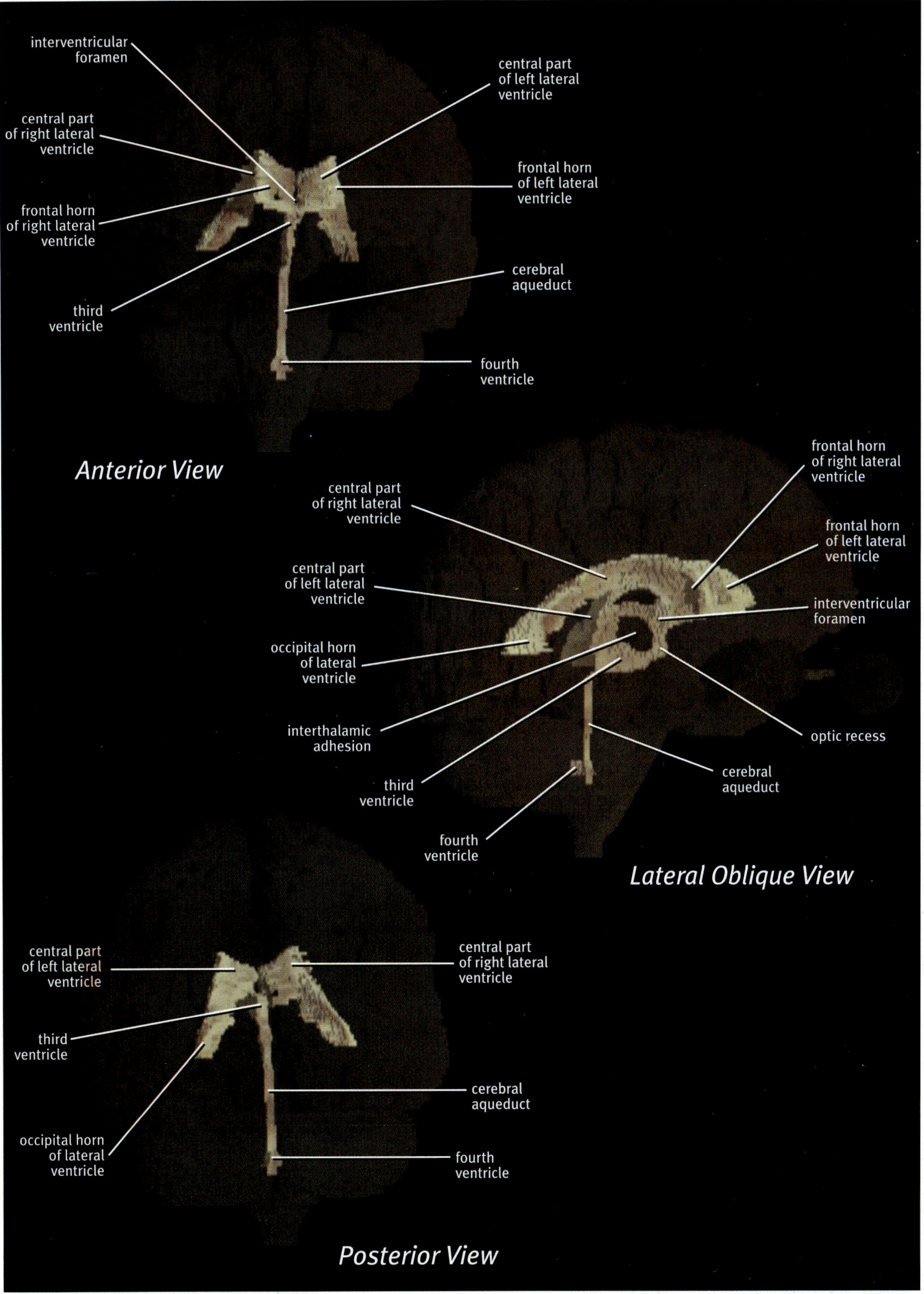

Cerebellum

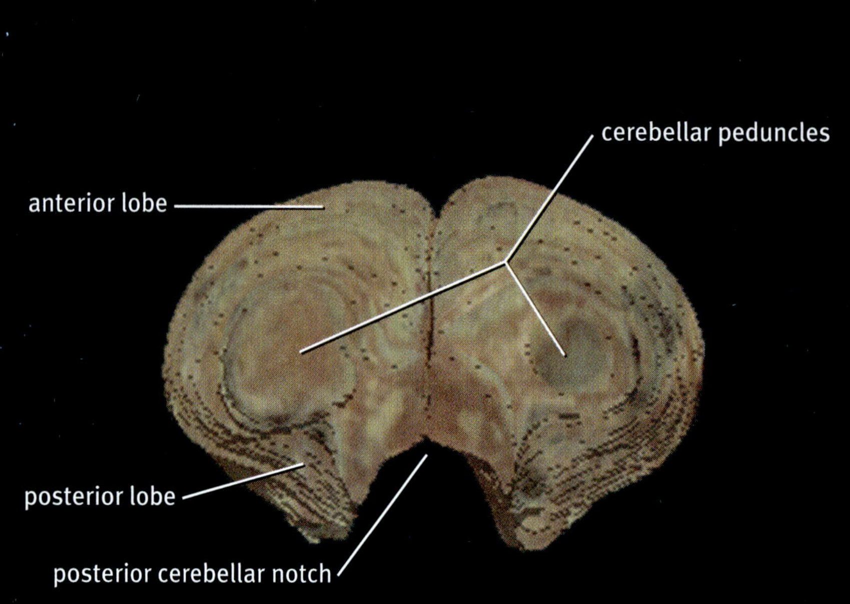

Inferior View

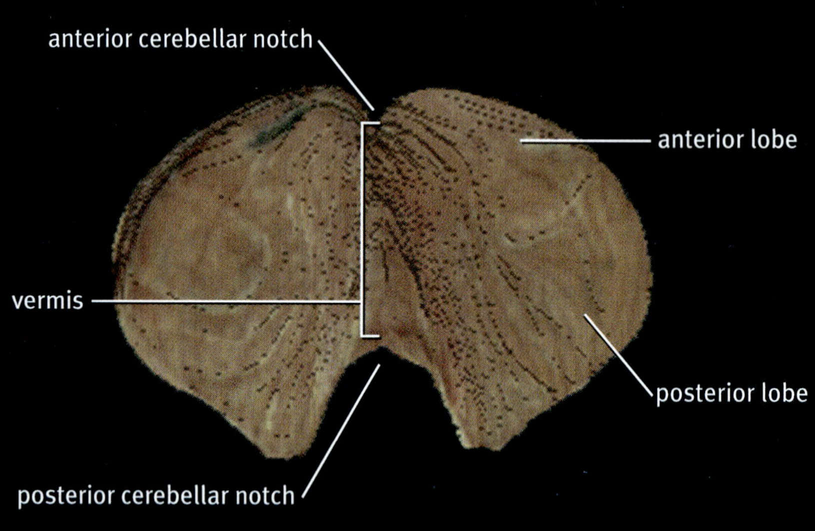

Superior View

Axial Sections of the Brain

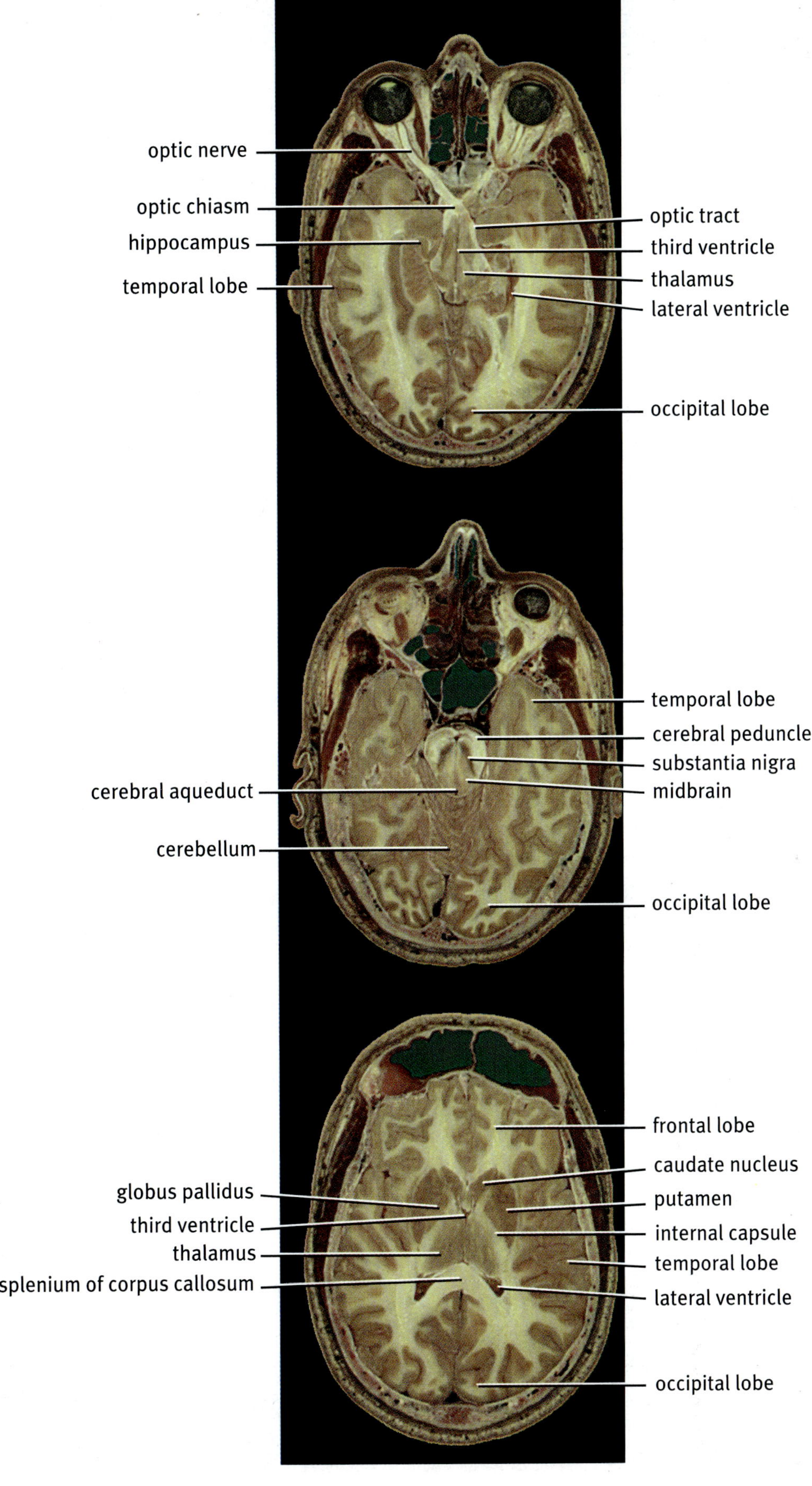

Sagittal Sections of the Brain

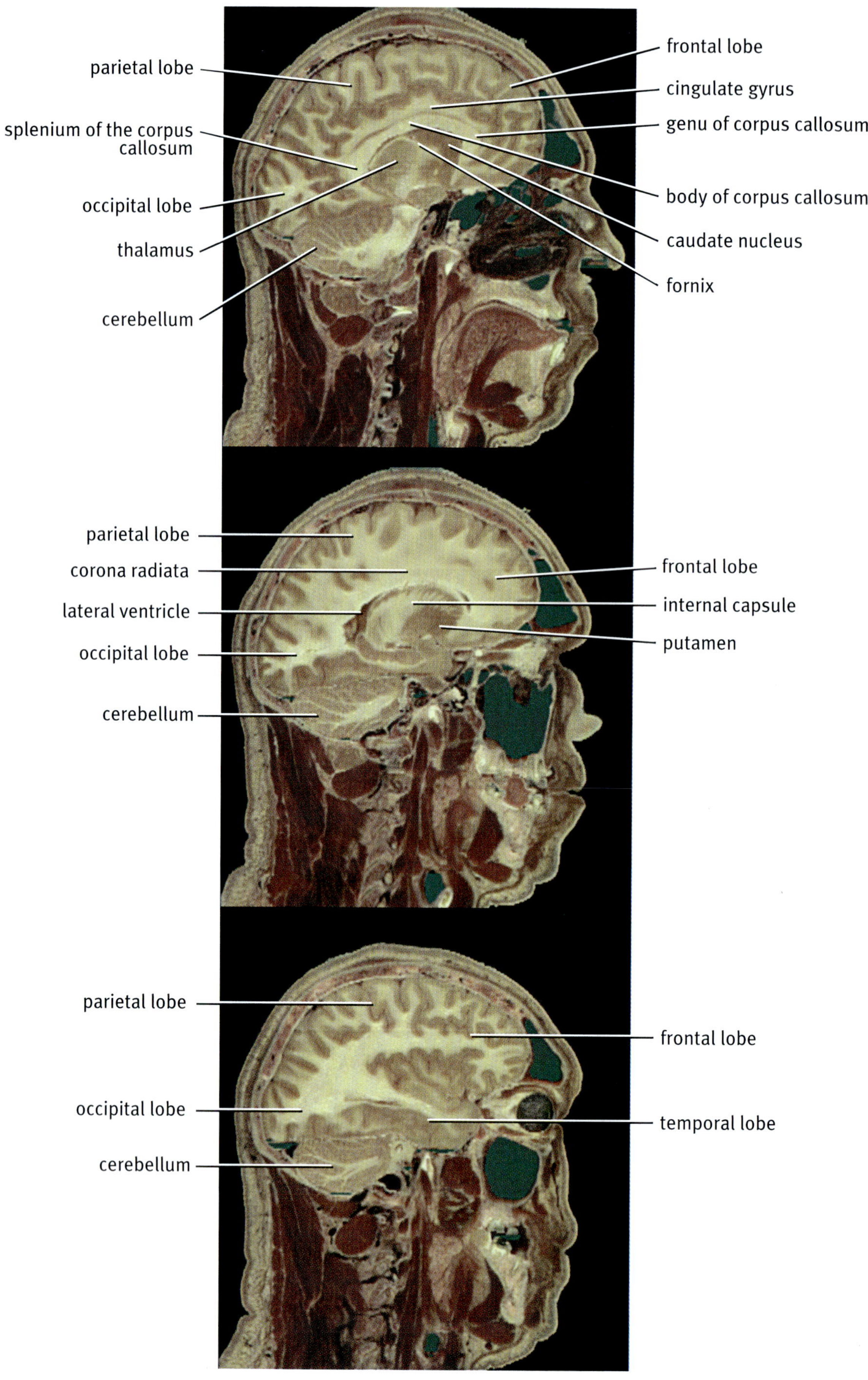

Coronal Sections of the Brain

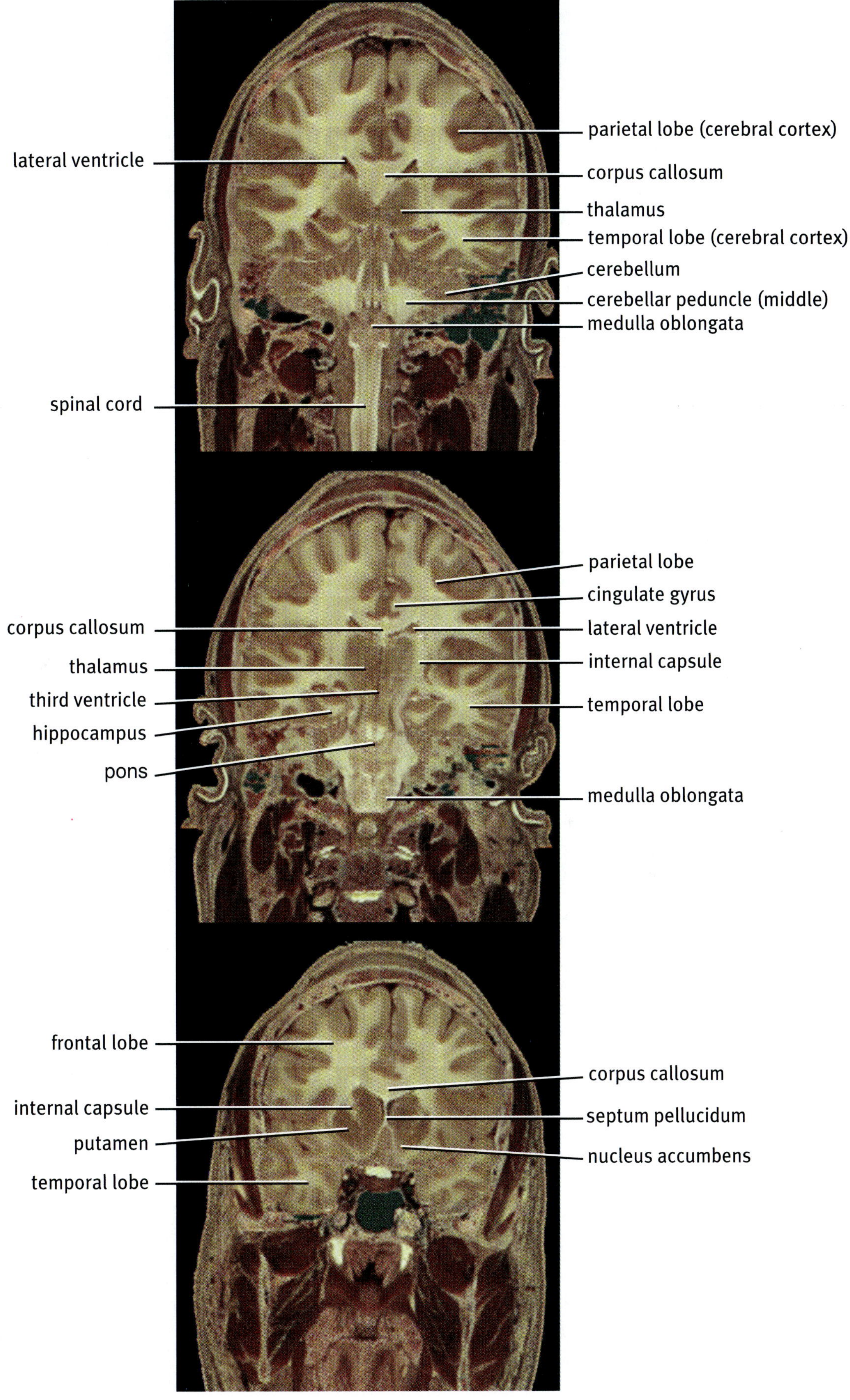

Brachial Plexus

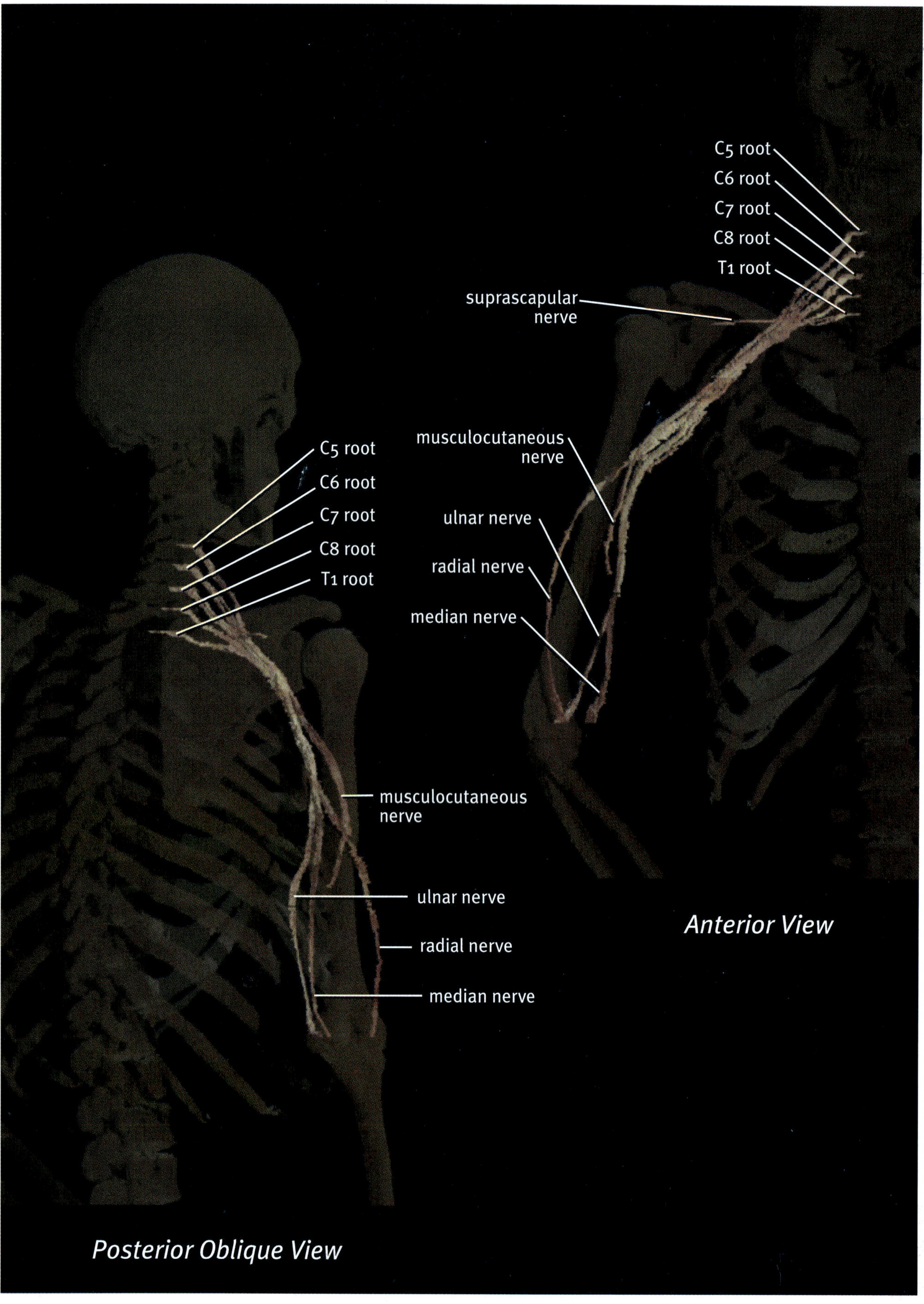

Lumbosacral Plexus

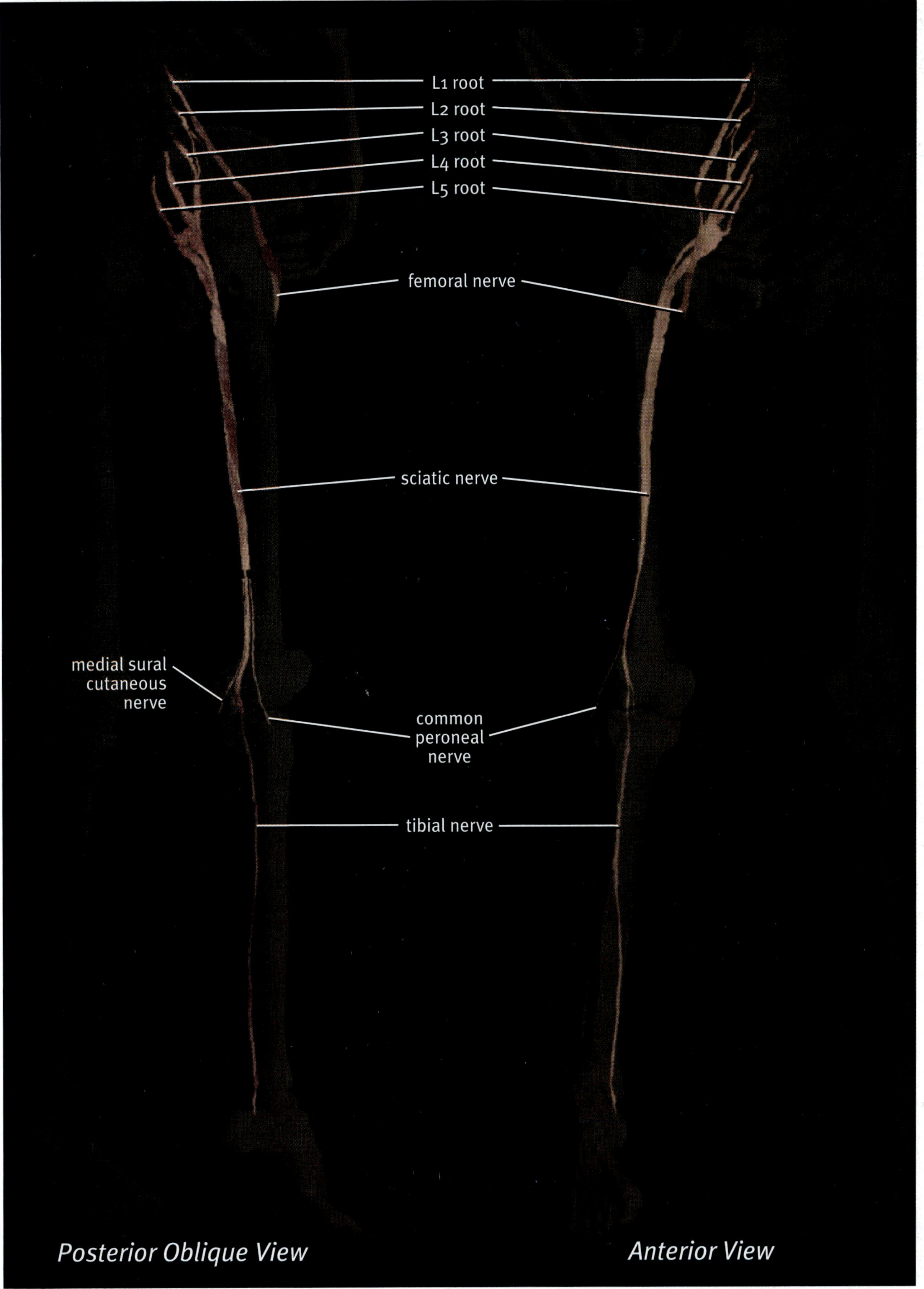

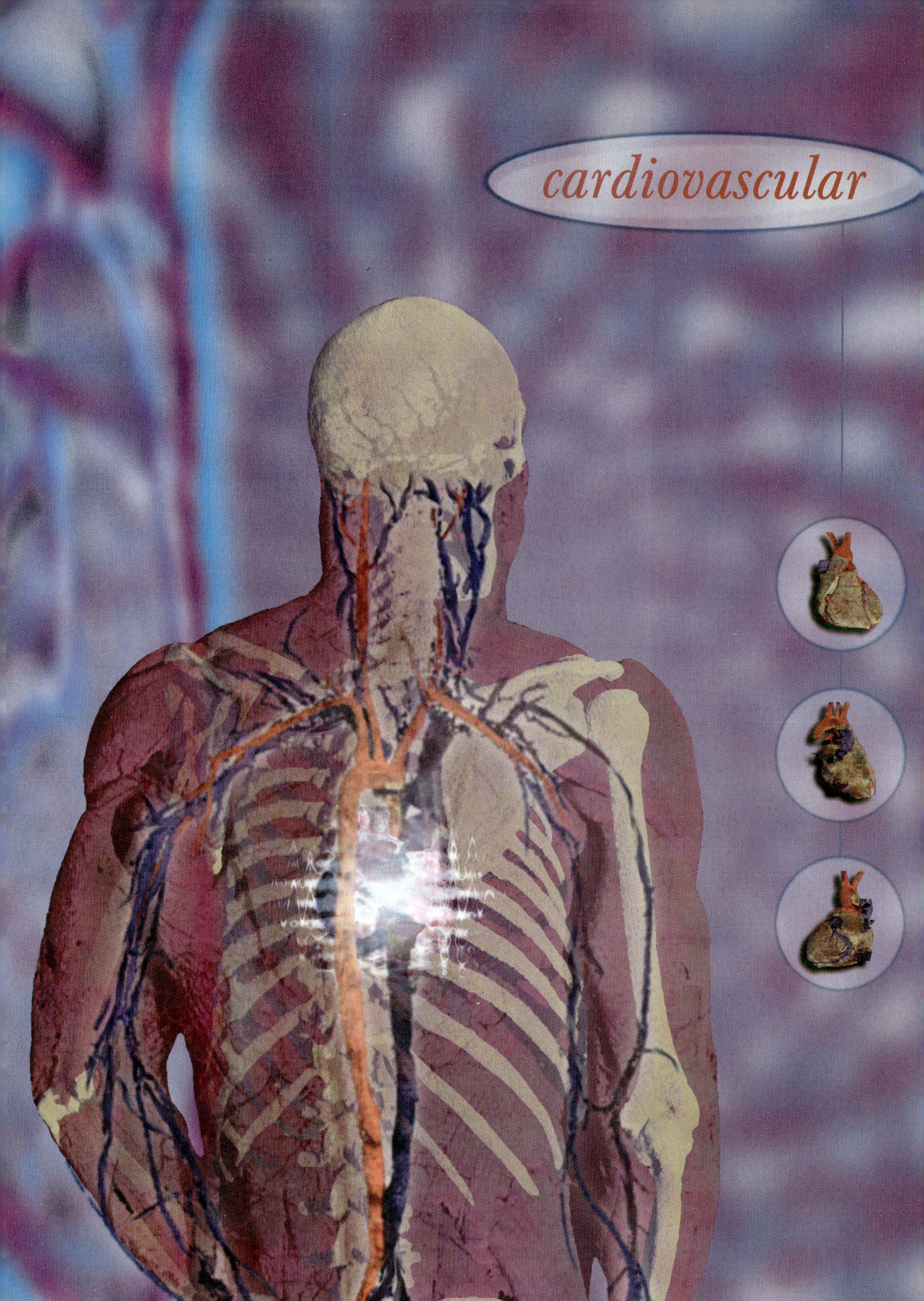
cardiovascular

Cardiovascular System 1

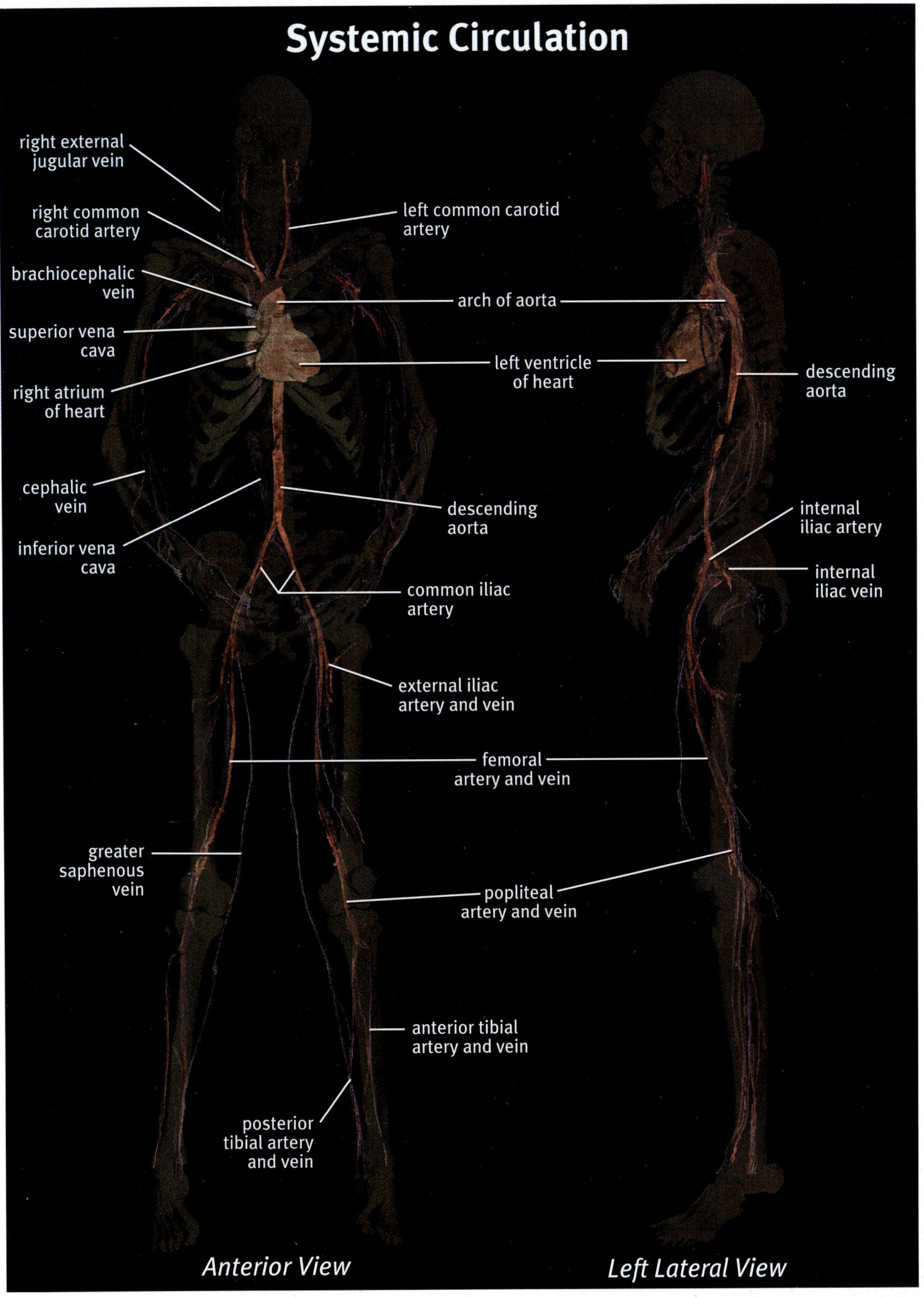

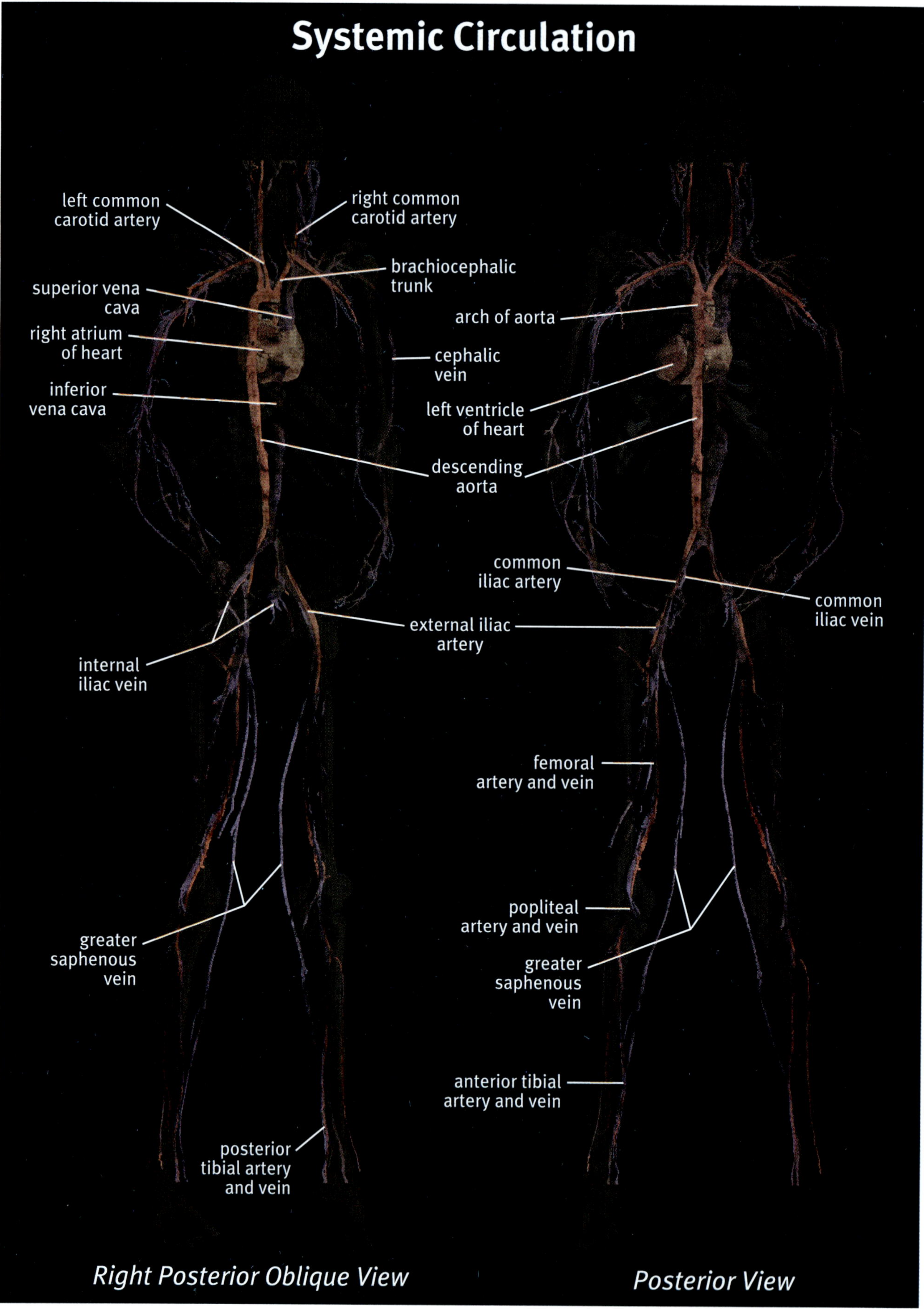
Systemic Circulation
left common carotid artery
right common carotid artery
brachiocephalic trunk
superior vena cava
right atrium of heart
arch of aorta
cephalic vein
inferior vena cava
left ventricle of heart
descending aorta
common iliac artery
common iliac vein
external iliac artery
internal iliac vein
femoral artery and vein
popliteal artery and vein
greater saphenous vein
greater saphenous vein
anterior tibial artery and vein
posterior tibial artery and vein
Right Posterior Oblique View
Posterior View

Heart

Left Posterior Oblique View
right pulmonary artery
coronary sulcus
Anterior View
(sternocostal surface)
right auricle
superior vena cava
coronary sulcus
right atrium
right ventricle
anterior interventricular sulcus and anterior interventricular artery
left ventricle
Posterior View
right pulmonary artery
left pulmonary artery
left atrium
coronary sinus
left ventricle
posterior interventricular sulcus and posterior descending artery
Superior View
inferior vena cava
arch of aorta
left subclavian artery
brachiocephalic trunk
left common carotid artery
right ventricle
left ventricle
Right Posterior Oblique View
brachiocephalic trunk
left common carotid artery
left subclavian artery
arch of aorta
superior vena cava
right pulmonary artery
right superior pulmonary vein
left superior pulmonary vein
left inferior pulmonary vein
right inferior pulmonary vein
inferior vena cava
Inferior View
anterior interventricular sulcus and anterior interventricular artery
right ventricle
left ventricle
right atrium
inferior vena cava
descending aorta

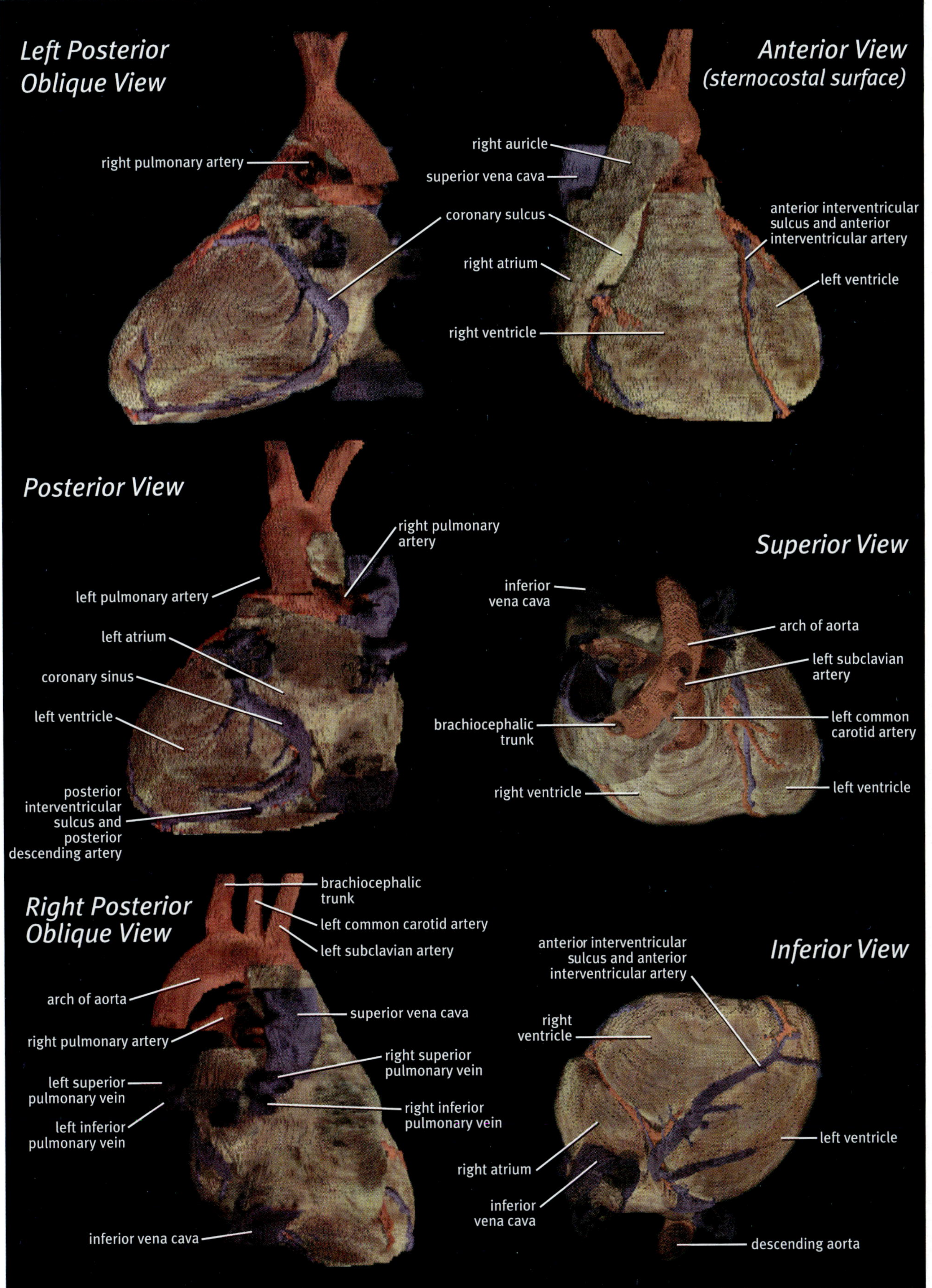

Sections Through the Heart 1

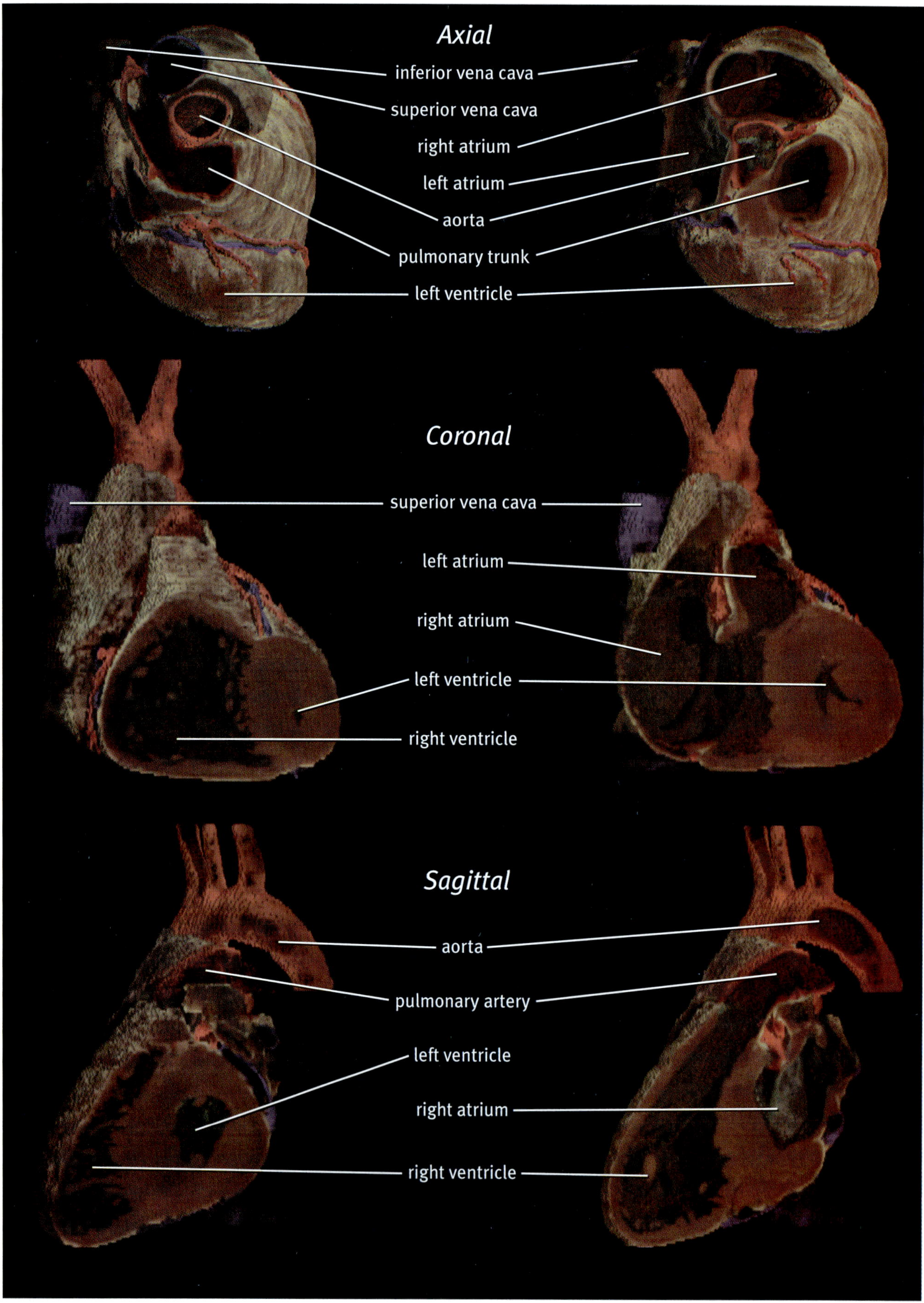

Sections Through the Heart 2

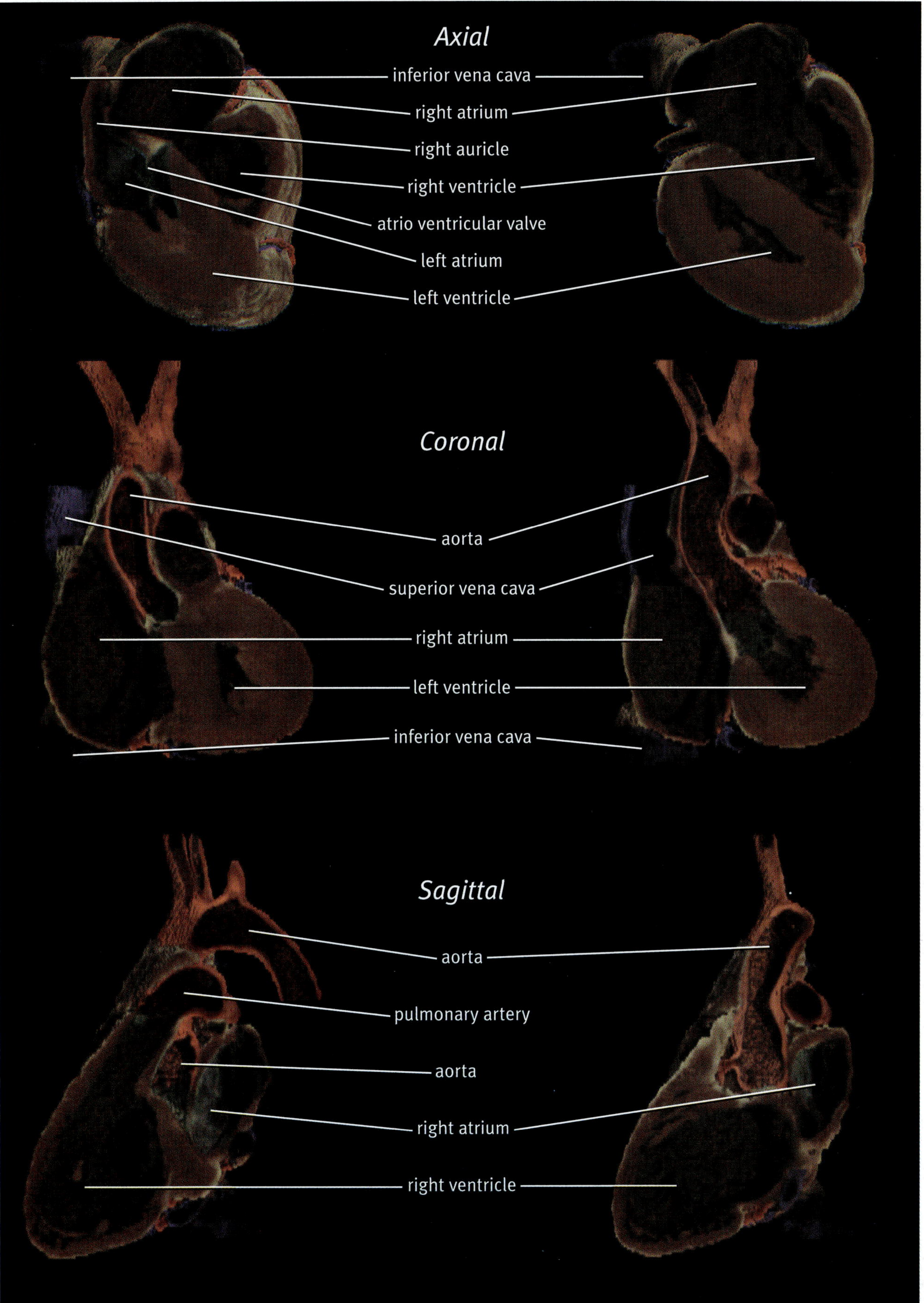

Cross-sections of the Heart 1

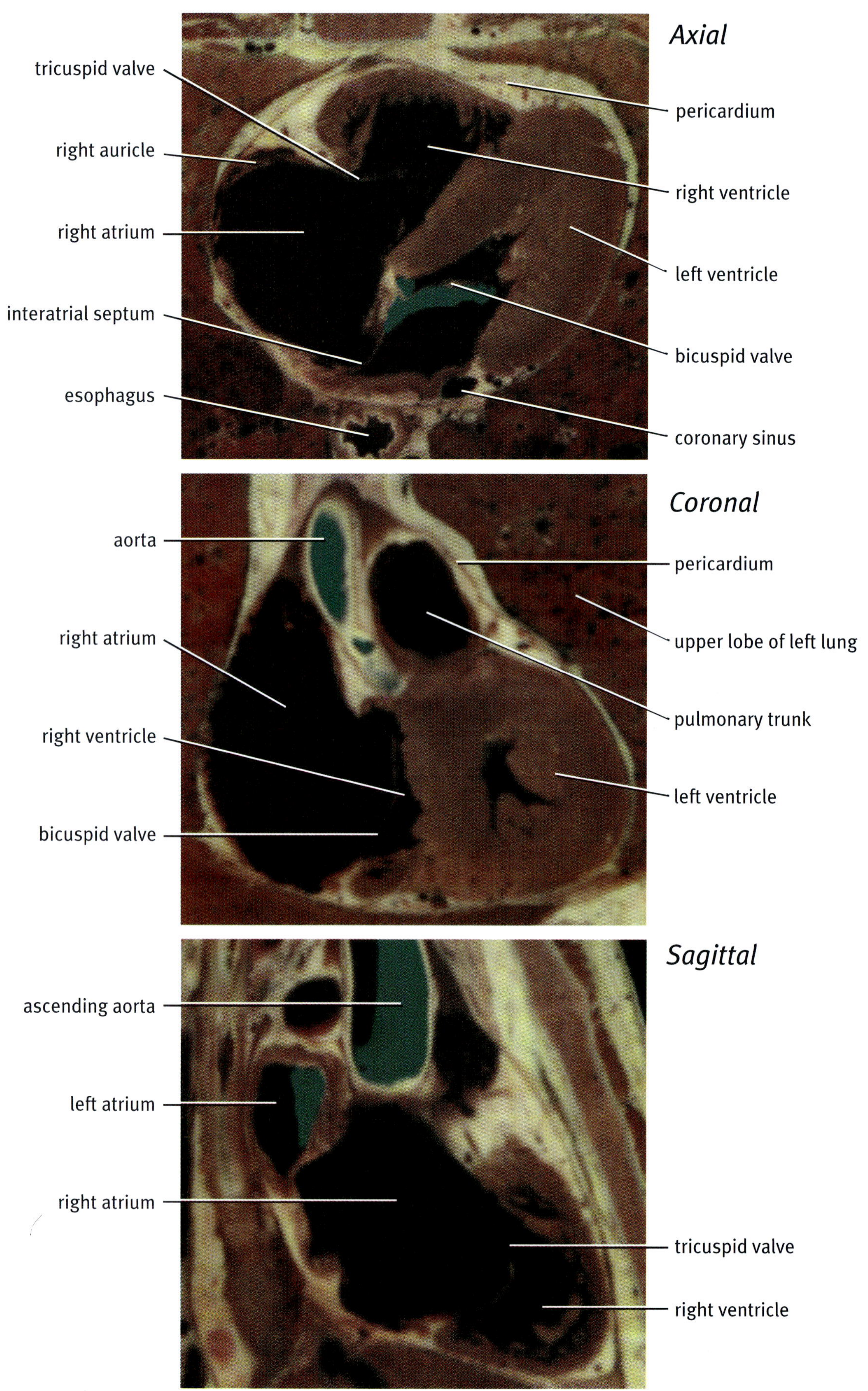

Cross-sections of the Heart 2

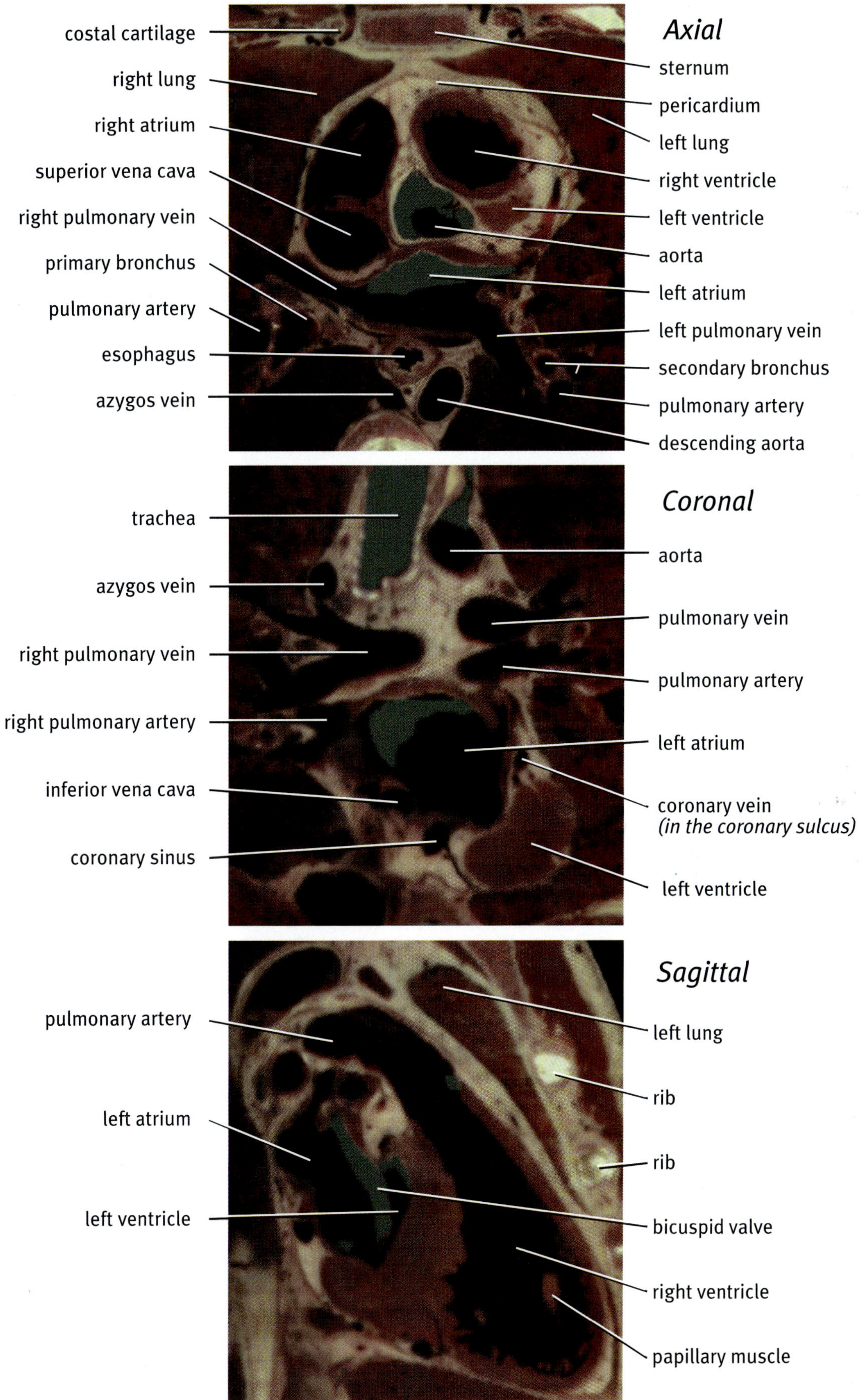

Vessels of the Head and Neck 1

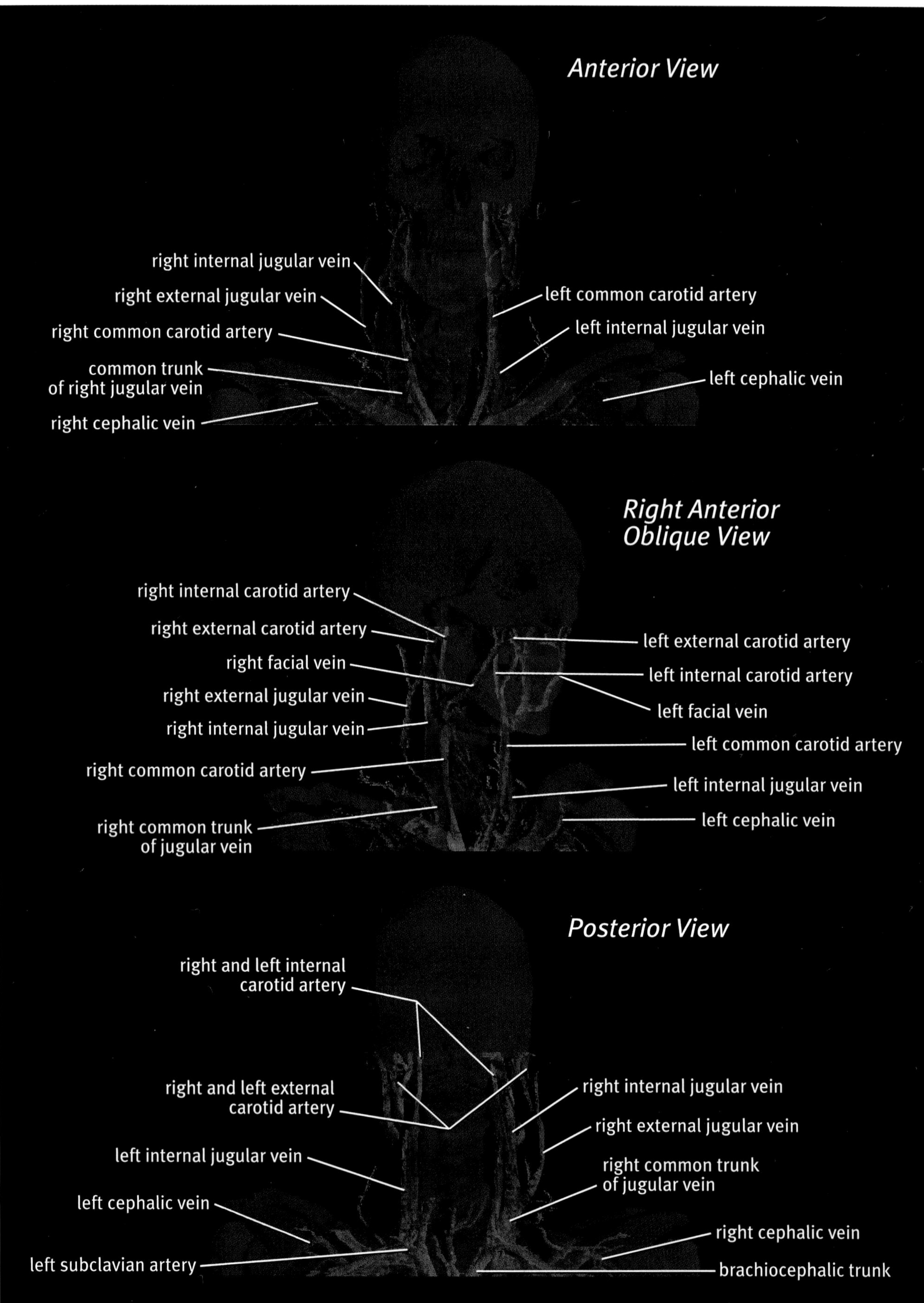

Vessels of the Head and Neck 2

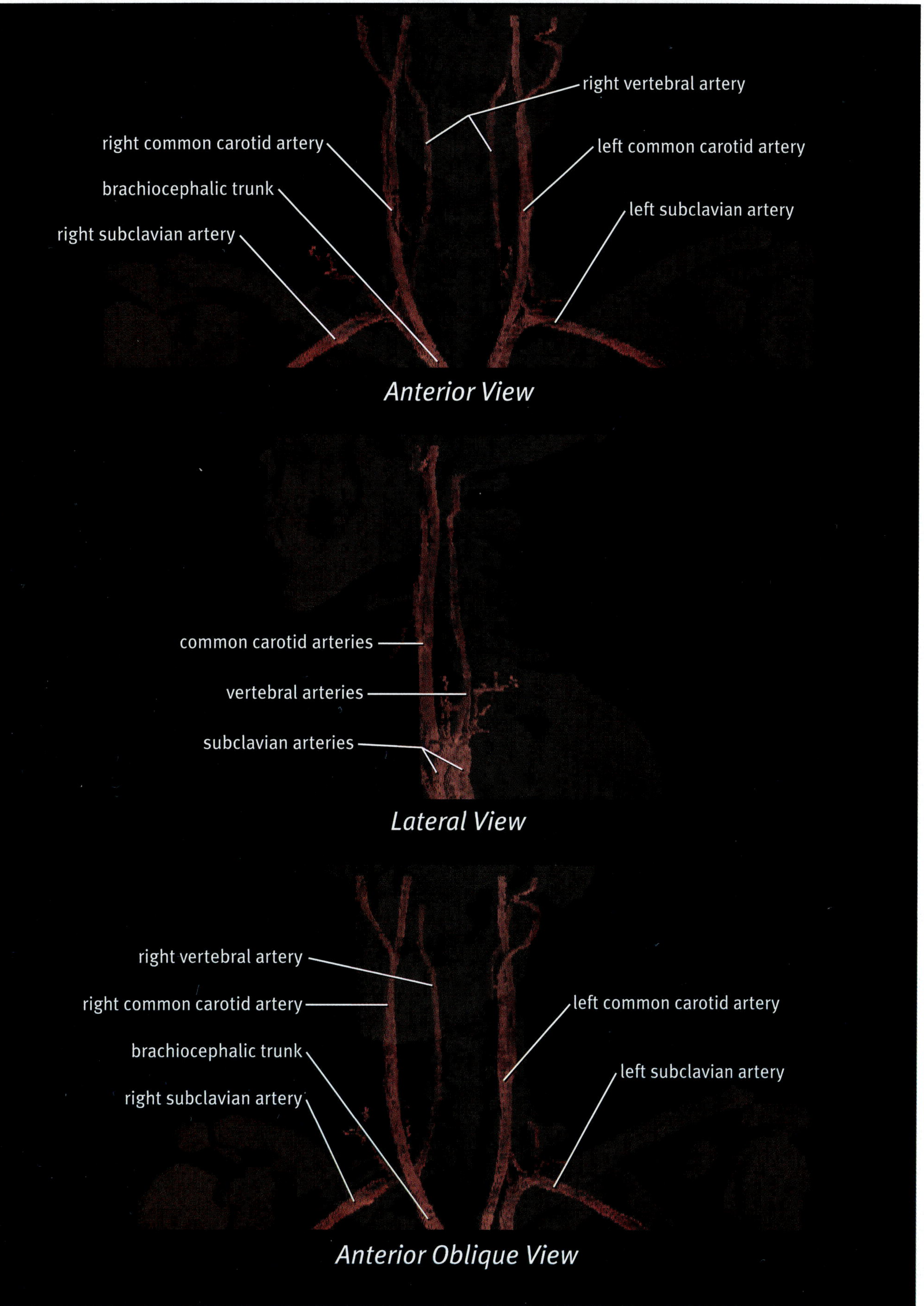

Great Vessels

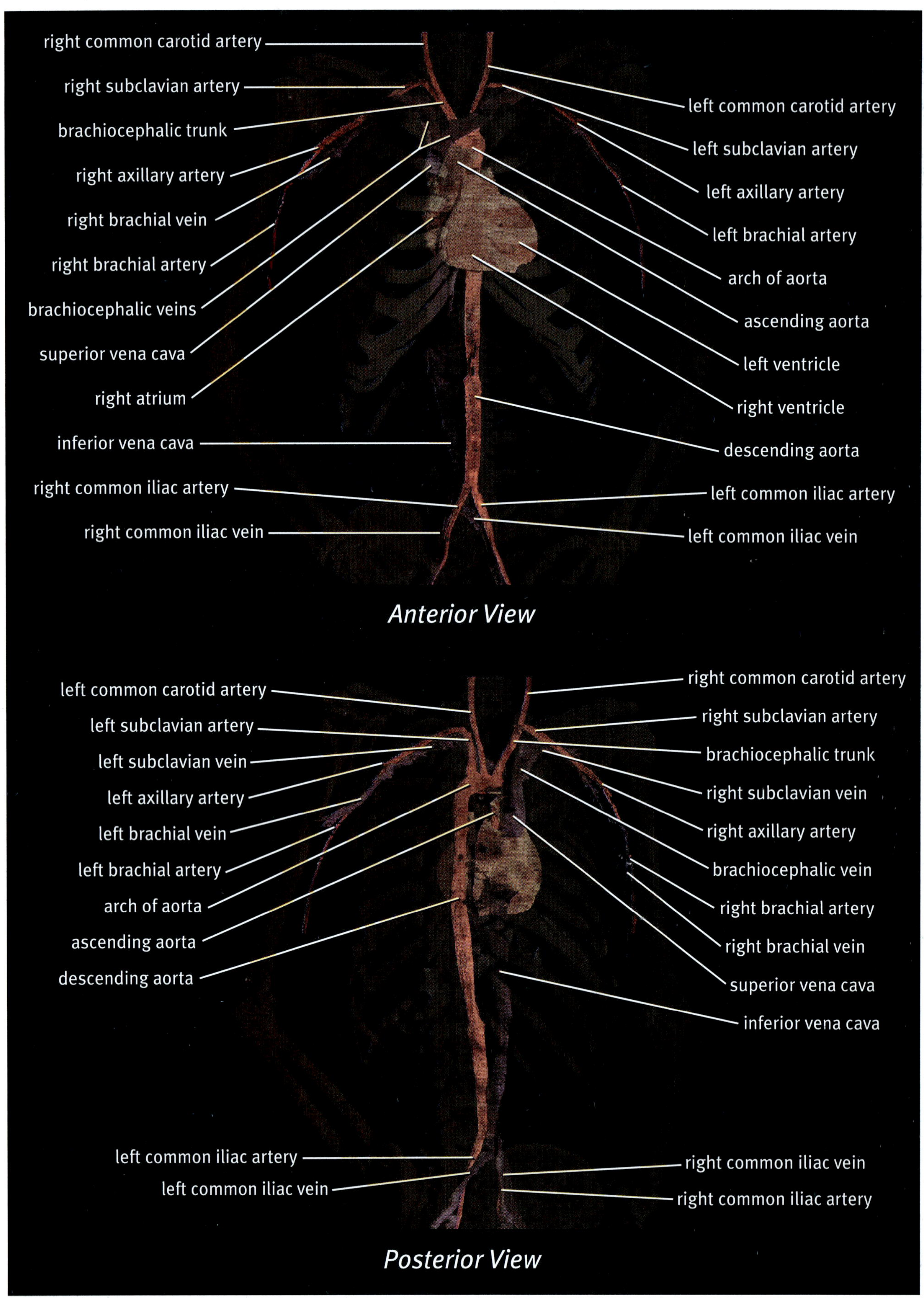

Abdominal Vasculature

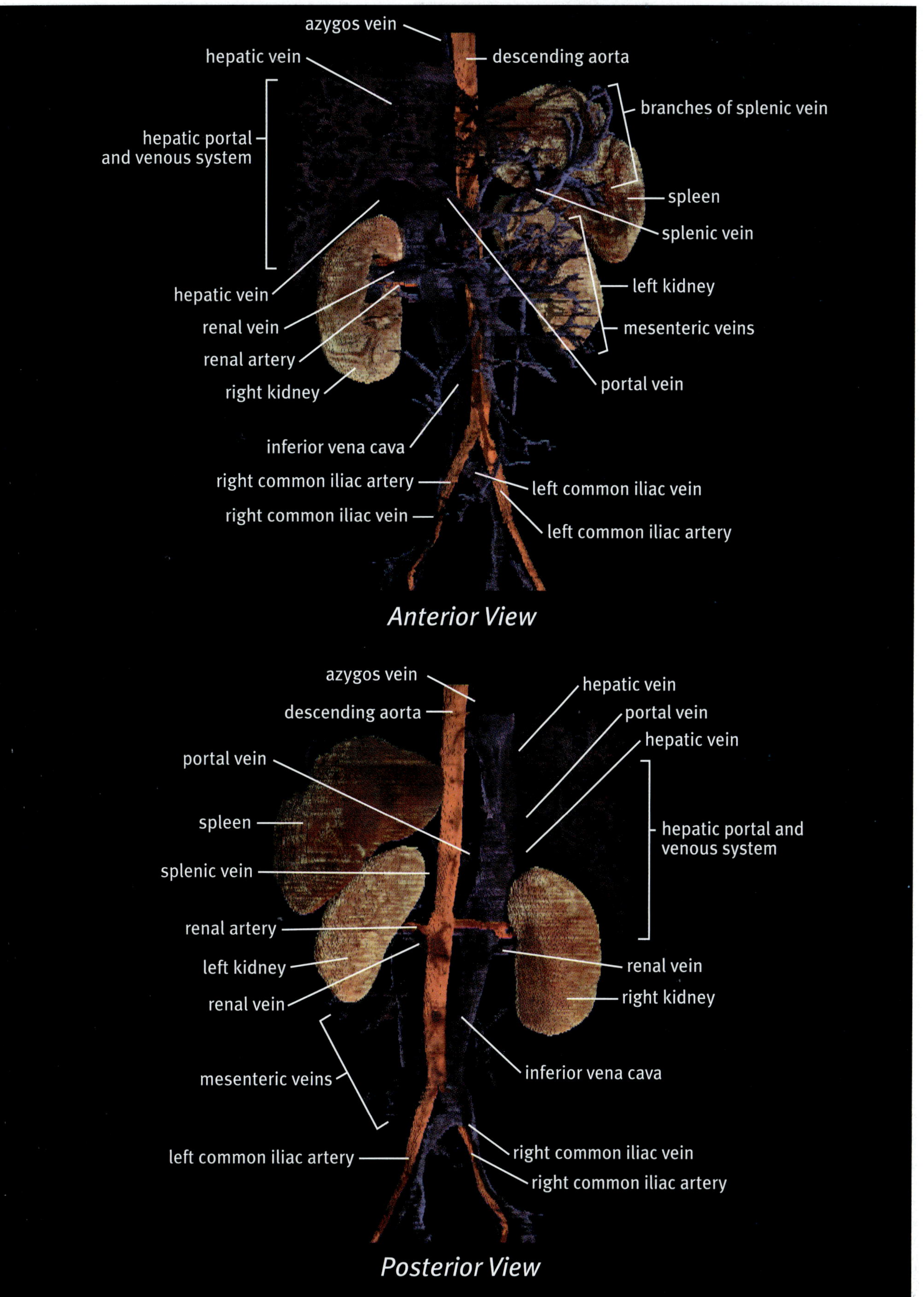

Anterior View

Posterior View

Bronchial Arteries and Veins 1

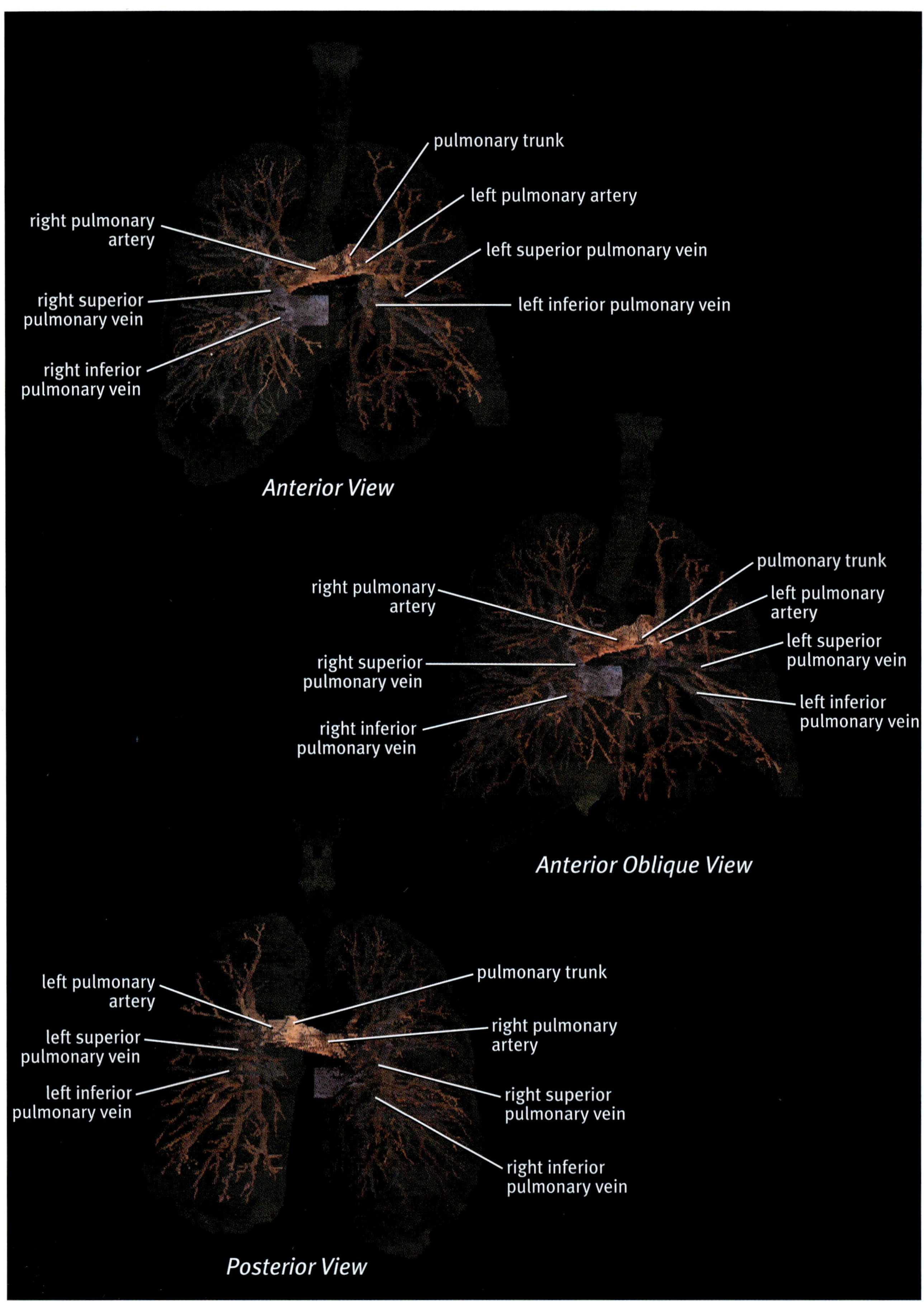

Bronchial Arteries and Veins 2

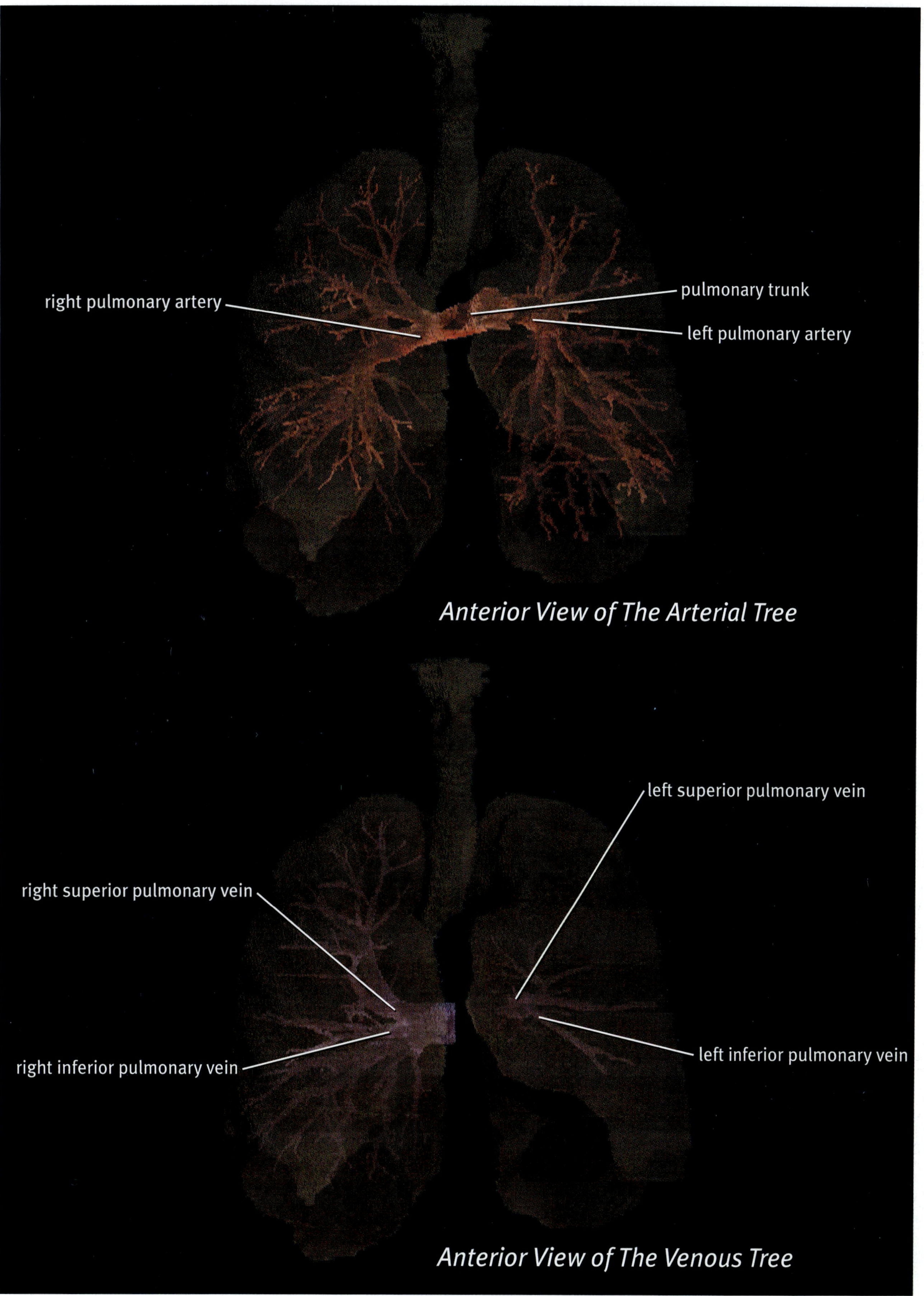

Anterior View of The Arterial Tree

Anterior View of The Venous Tree

Hepatic Portal System 1

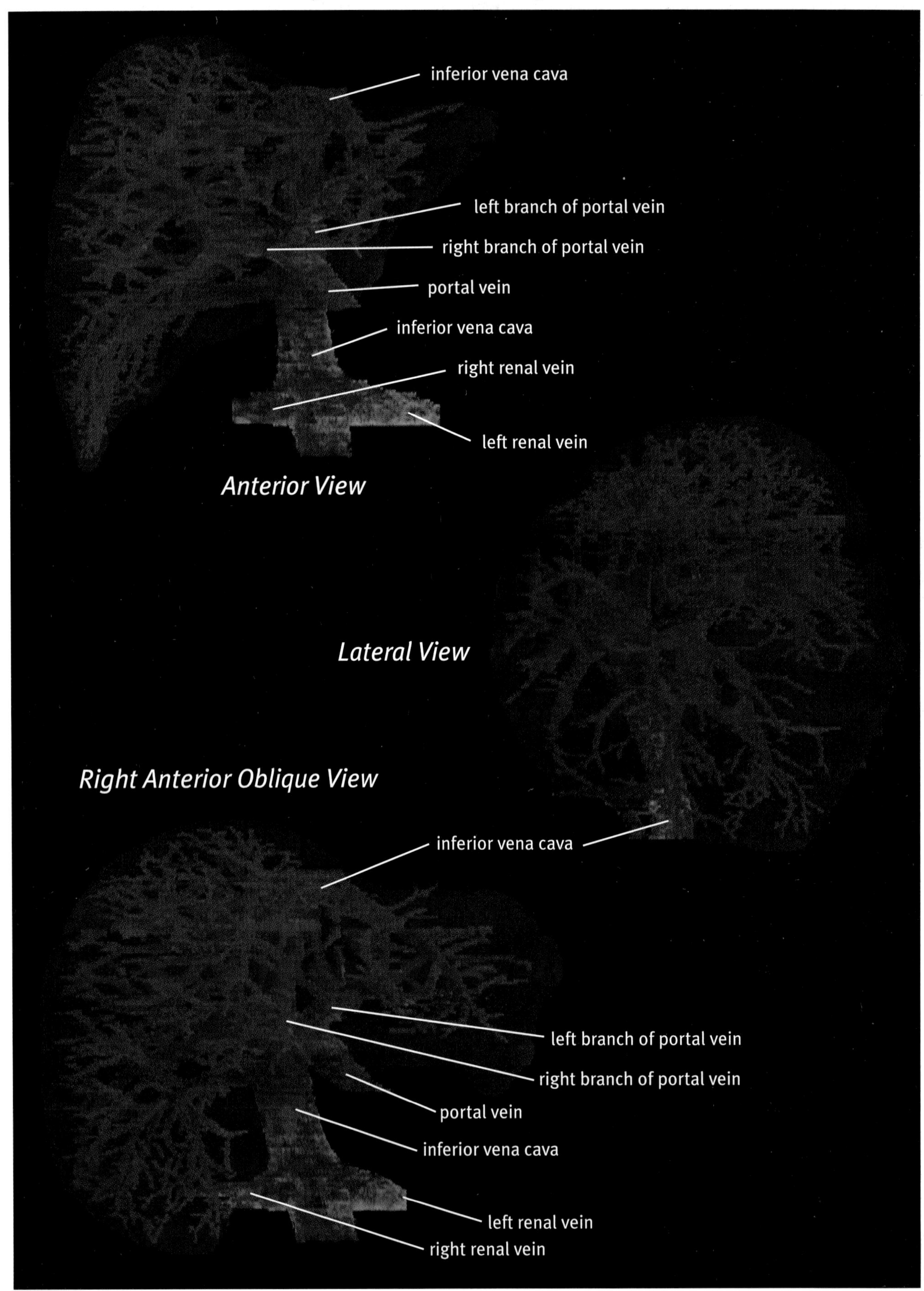

Hepatic Portal System 2

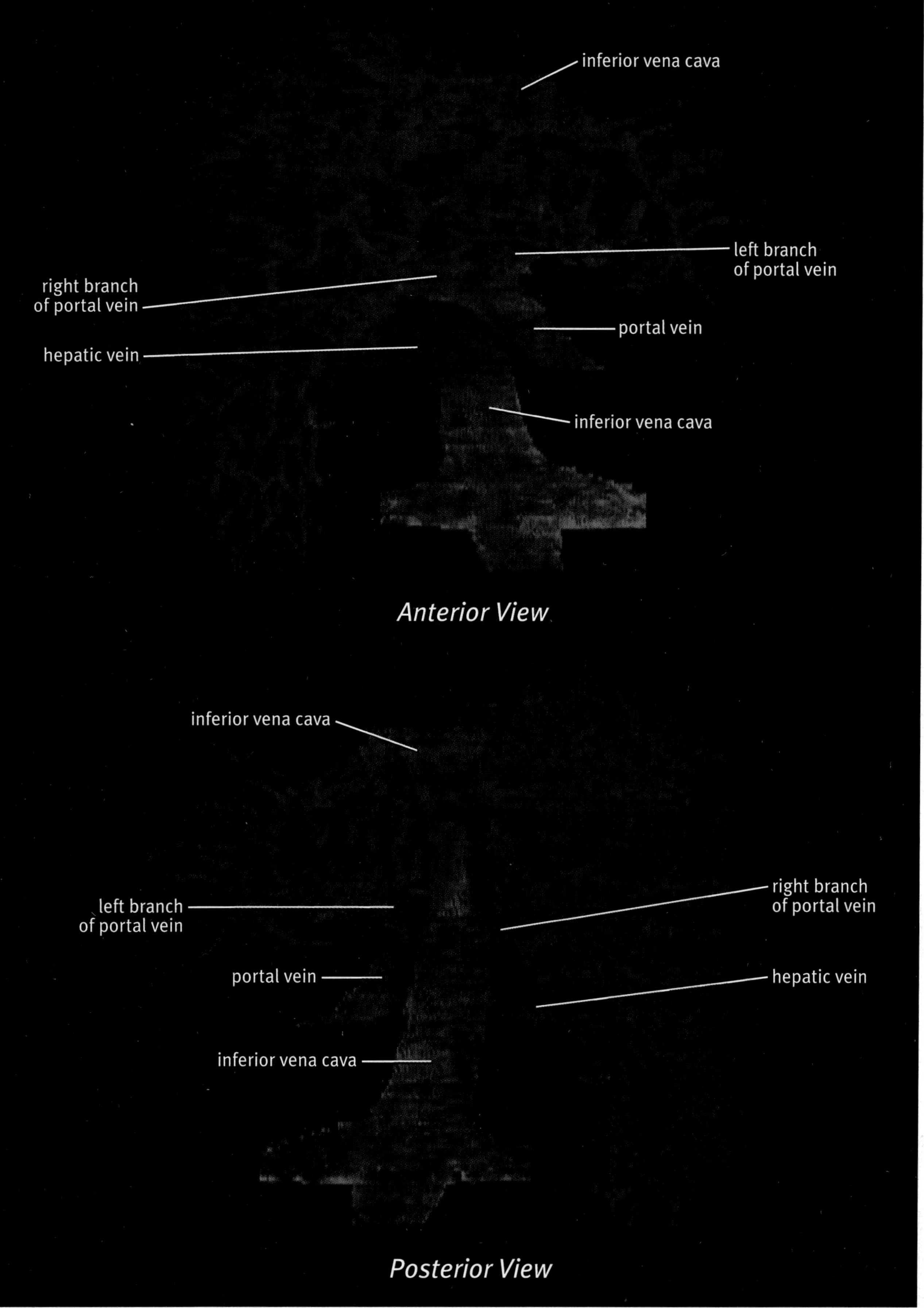

Anterior View

Posterior View

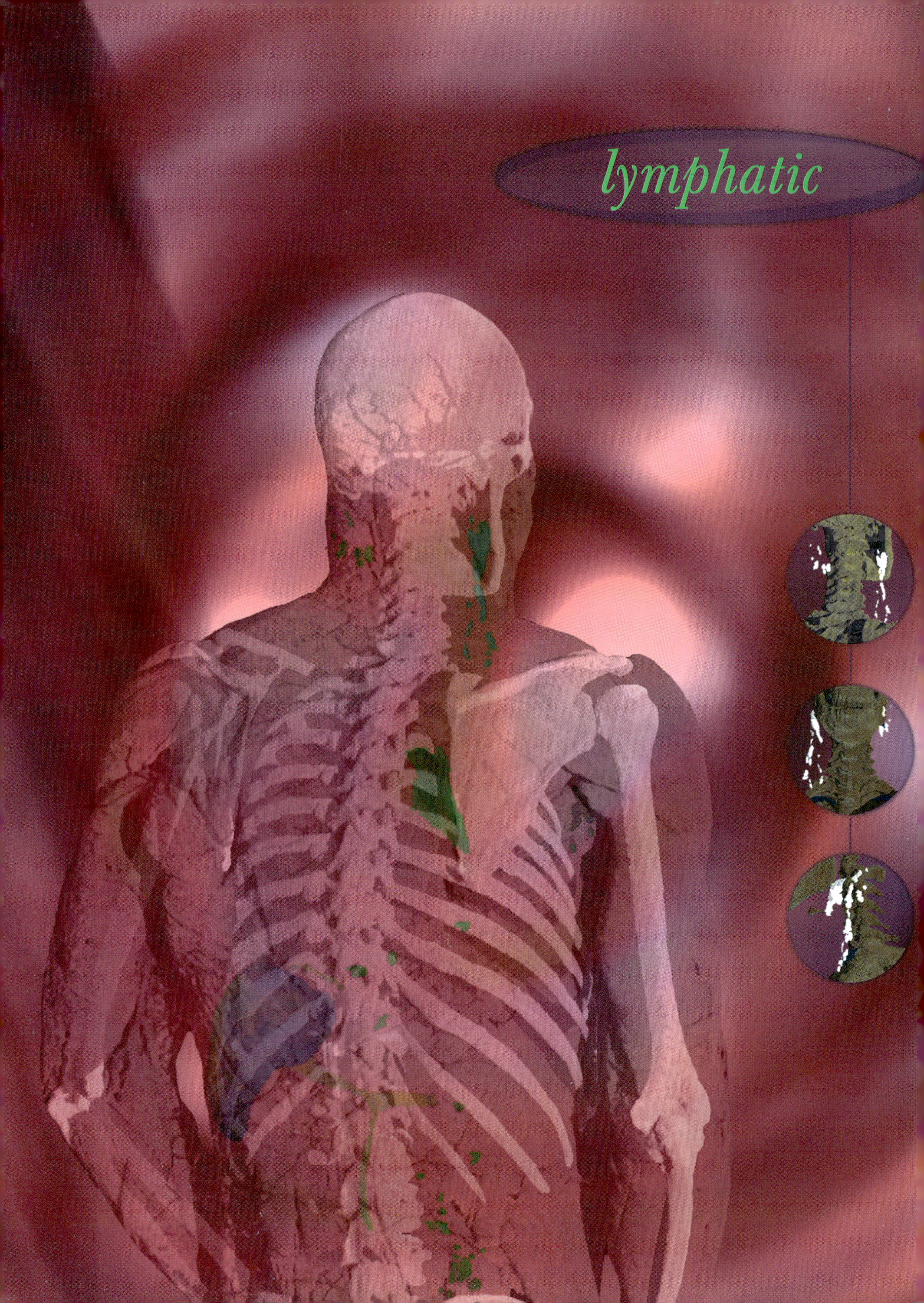
lymphatic

Lymphatic System

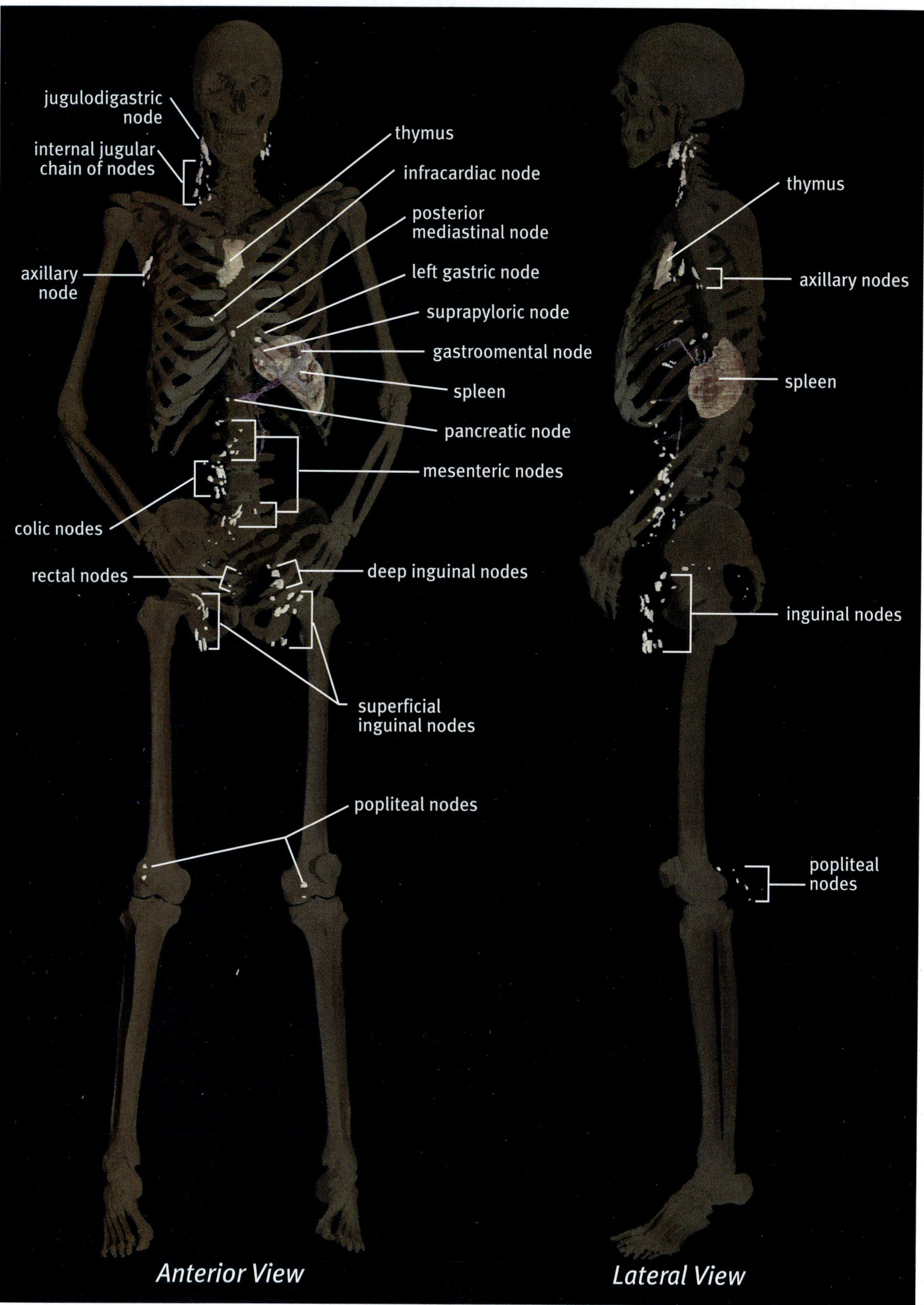

Spleen and Thymus

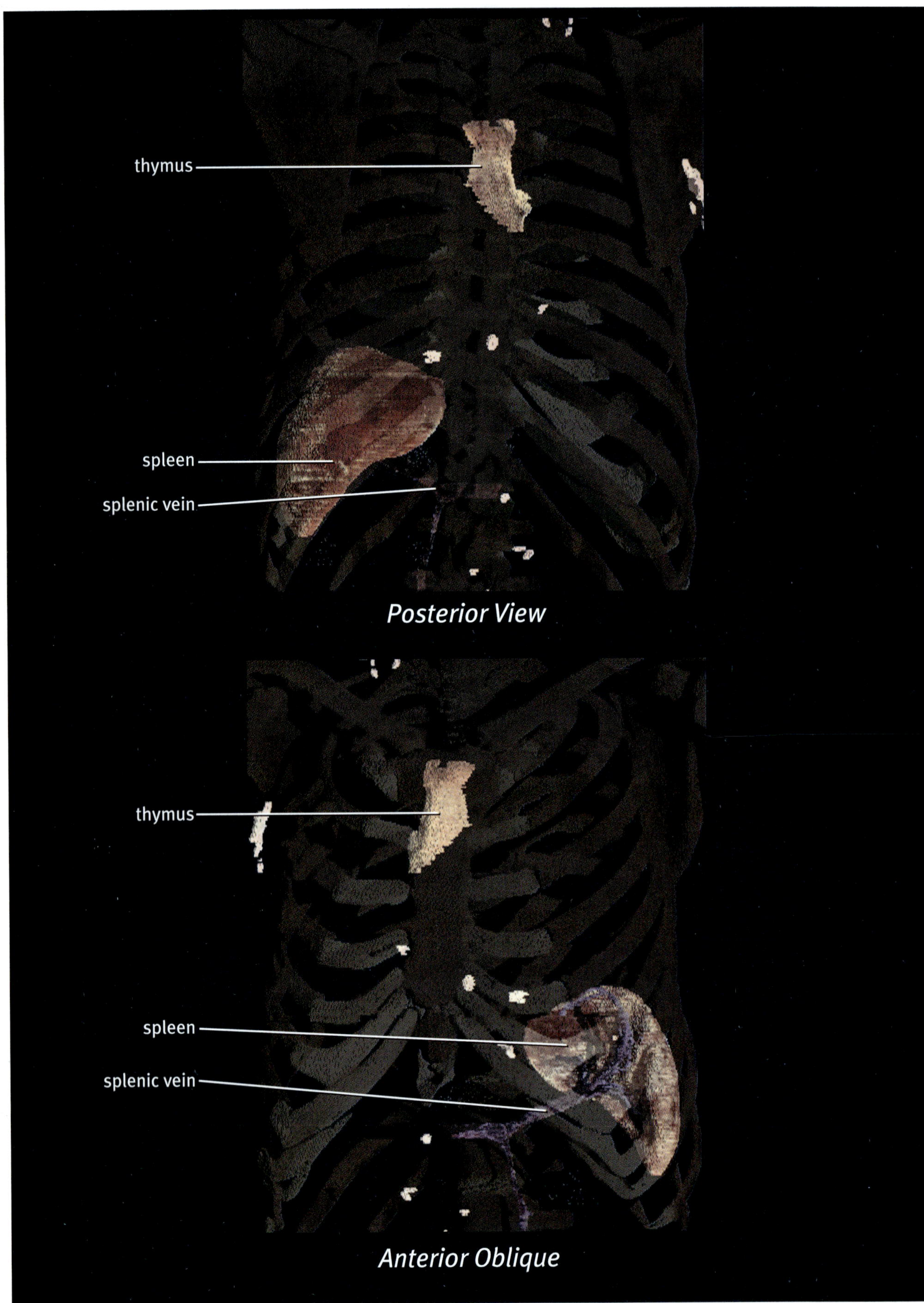

Posterior View

Anterior Oblique

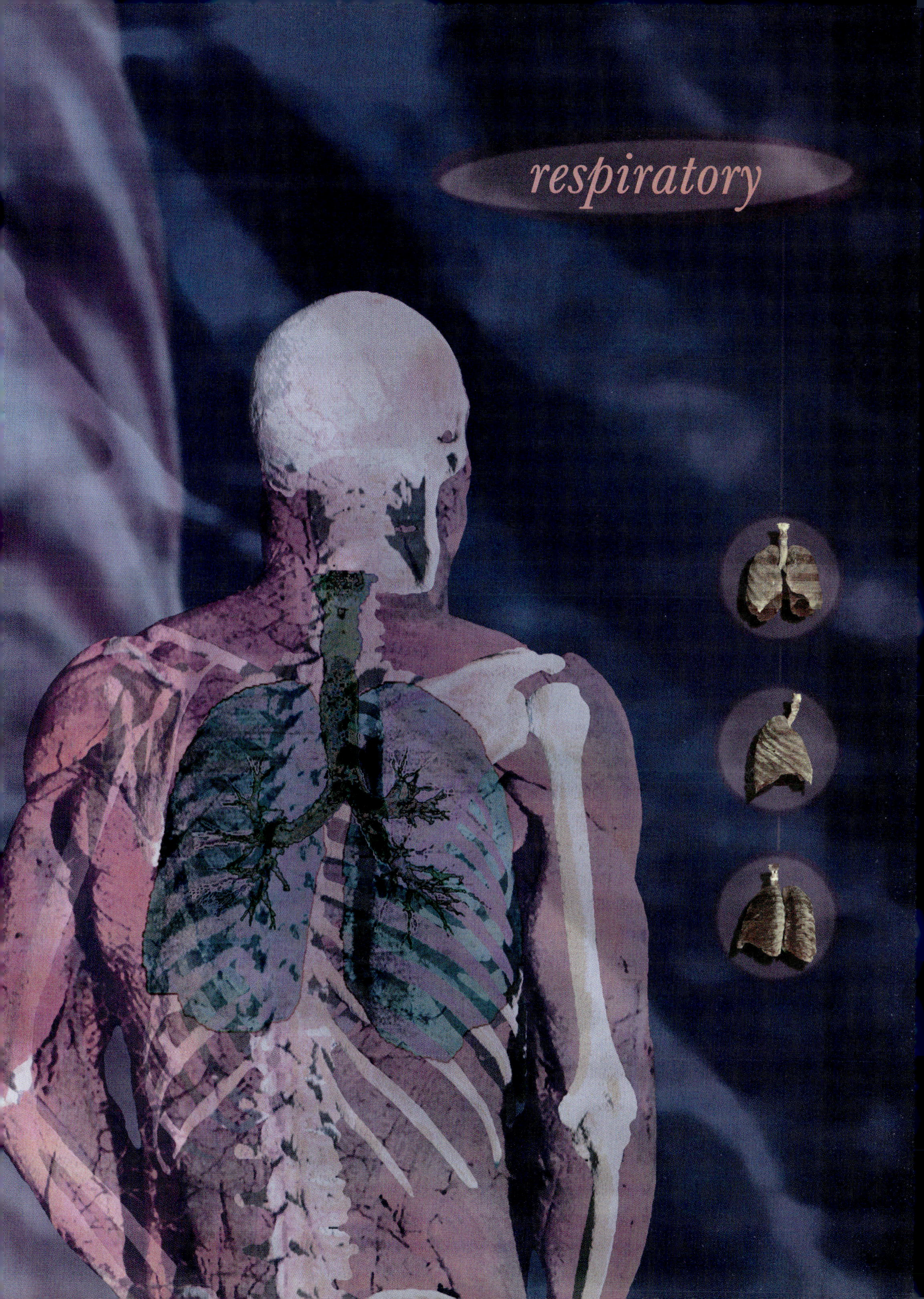
respiratory

Respiratory System 1

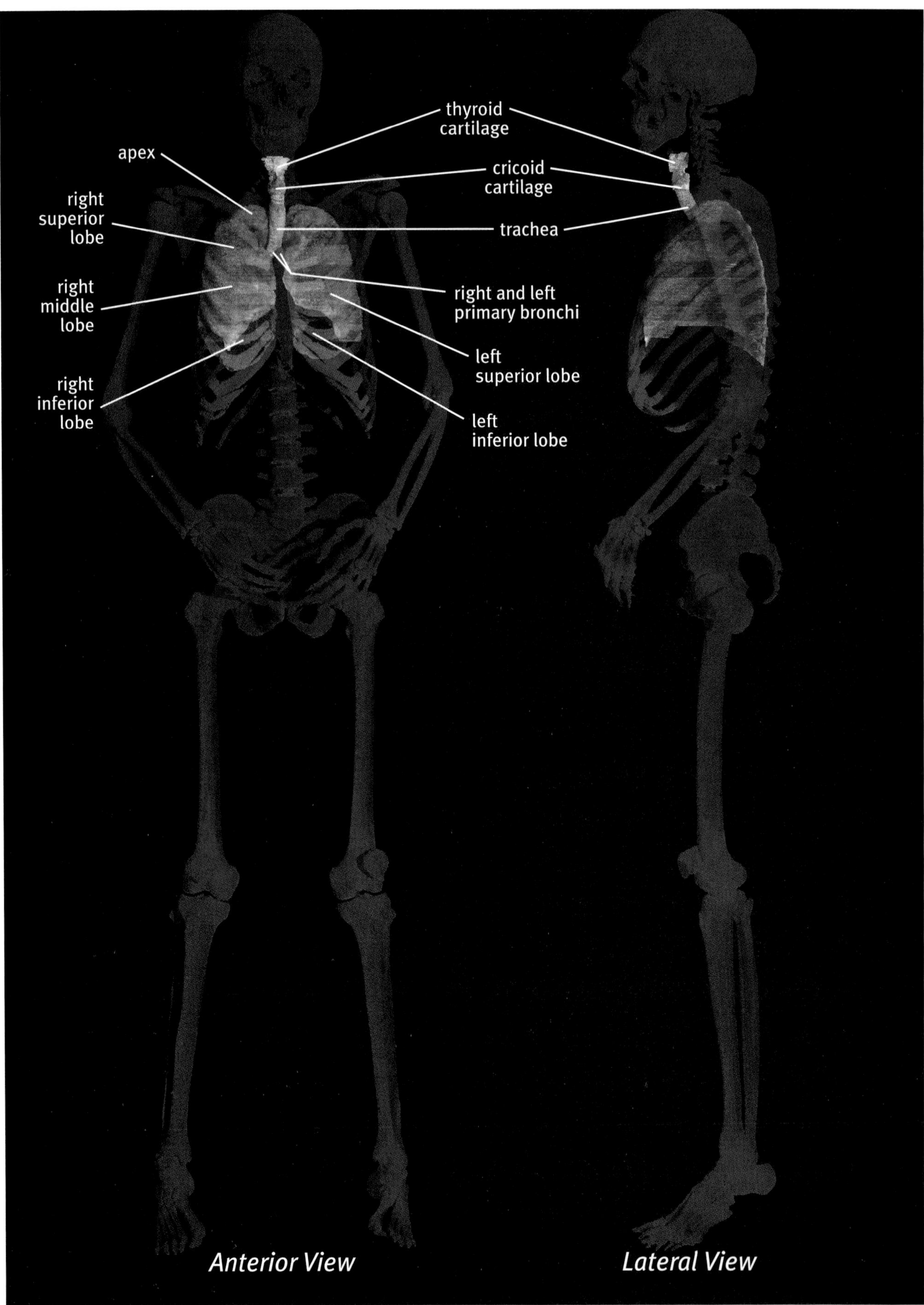

Respiratory System 2

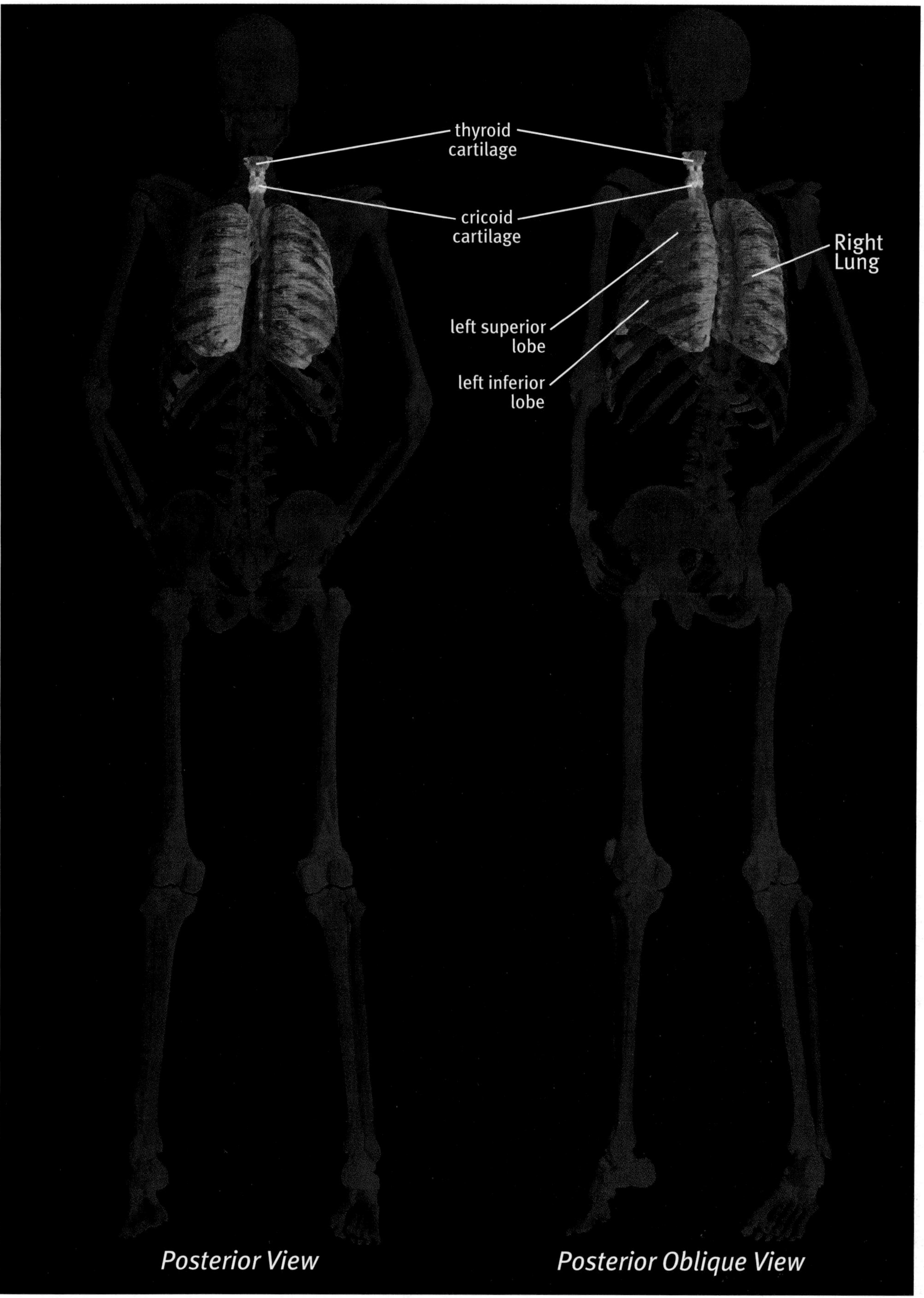

Lungs 1

thyroid cartilage
cricoid cartilage
apex
trachealis muscle
left superior lobe
right superior lobe
right oblique fissure
left oblique fissure
left inferior lobe
right inferior lobe
Posterior View
thyroid cartilage
cricoid cartilage
apex
trachea
right superior lobe
left superior lobe
right middle lobe
right and left primary bronchi
right inferior lobe
left inferior lobe
diaphragmatic surface
Anterior View
thyroid cartilage
cricoid cartilage
trachea
trachealis muscle
right superior lobe
left superior lobe
lingula
left inferior lobe
right inferior lobe
Left Posterior Oblique View

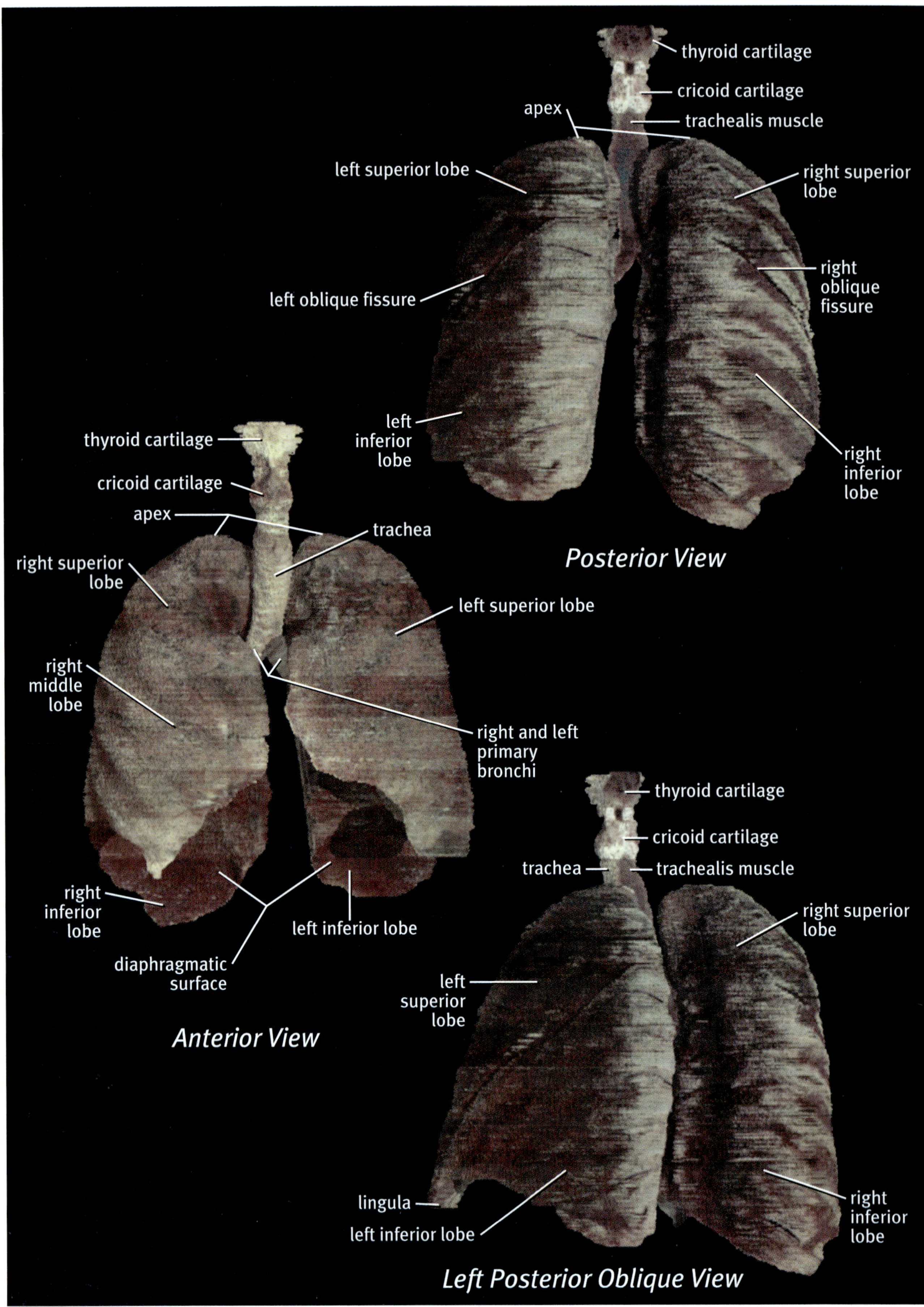

Lungs 2

thyroid cartilage
cricoid cartilage
trachea
apex
left superior lobe
left inferior lobe
left oblique fissure
lingular division
Lateral View of Left Lung
apex
impression for trachea and esophagus
oblique fissure
superior lobe
cardiac impression
inferior lobe
lingula
diaphragmatic surface
Medial View of Left Lung
thyroid cartilage
cricoid cartilage
apex
right superior lobe
right oblique fissure
horizontal fissure
right middle lobe
right inferior lobe
Lateral View of Right Lung
apex
impression for trachea
oblique fissure
superior lobe
inferior lobe
middle lobe
impression for esophagus
Medial View of Right Lung

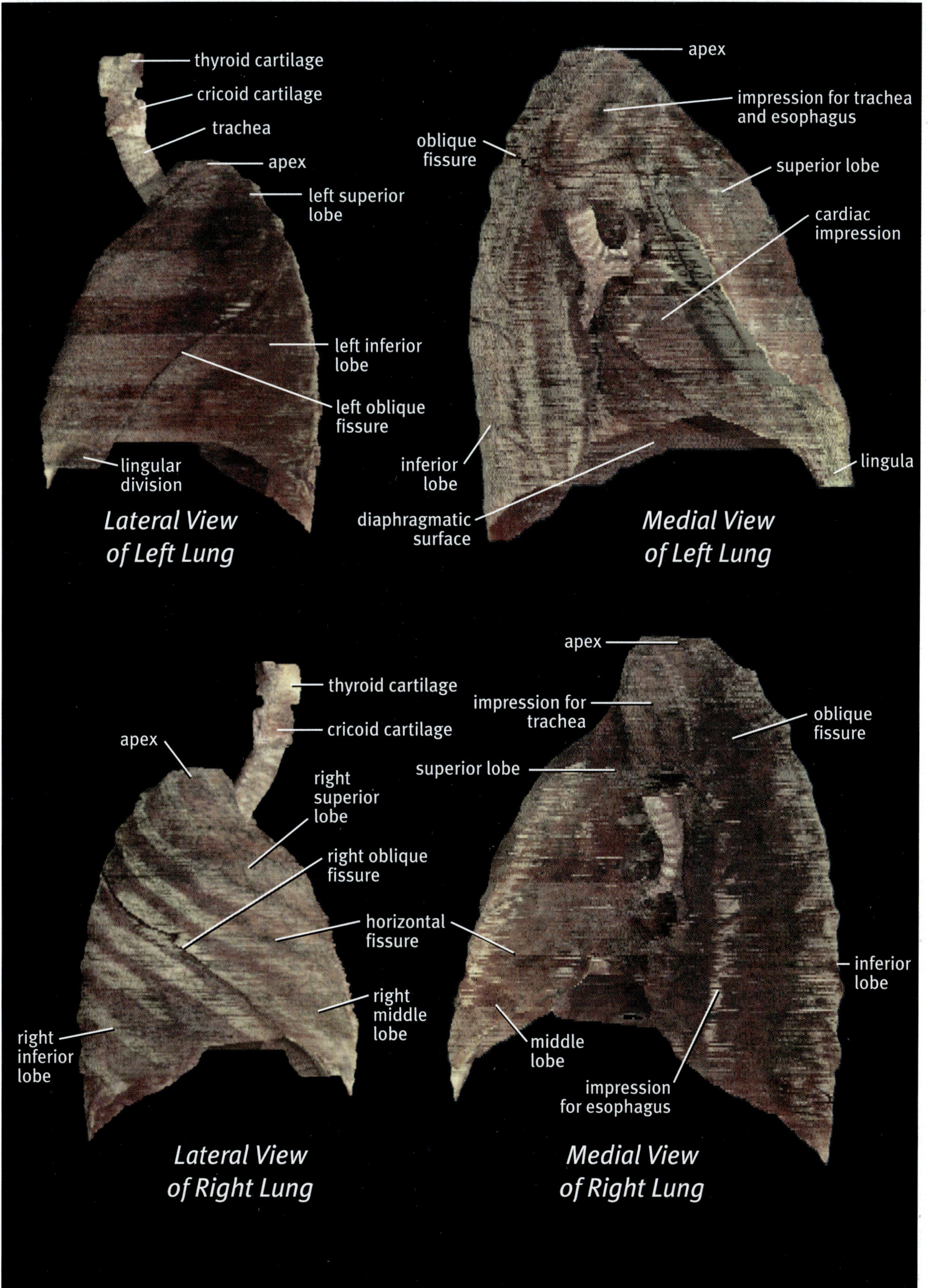

Airways

thyroid cartilage
cricoid cartilage
trachea
right superior lobe bronchus
left superior bronchus
left superior lobe bronchus
left lingular bronchus
left inferior lobe bronchus
right inferior lobe bronchus
right and left primary bronchi
Anterior View
thyroid cartilage
cricoid cartilage
trachealis muscle
left superior bronchus
right superior lobe bronchus
left superior lobe bronchus
left inferior lobe bronchus
right inferior lobe bronchus
right and left primary bronchi
Posterior View

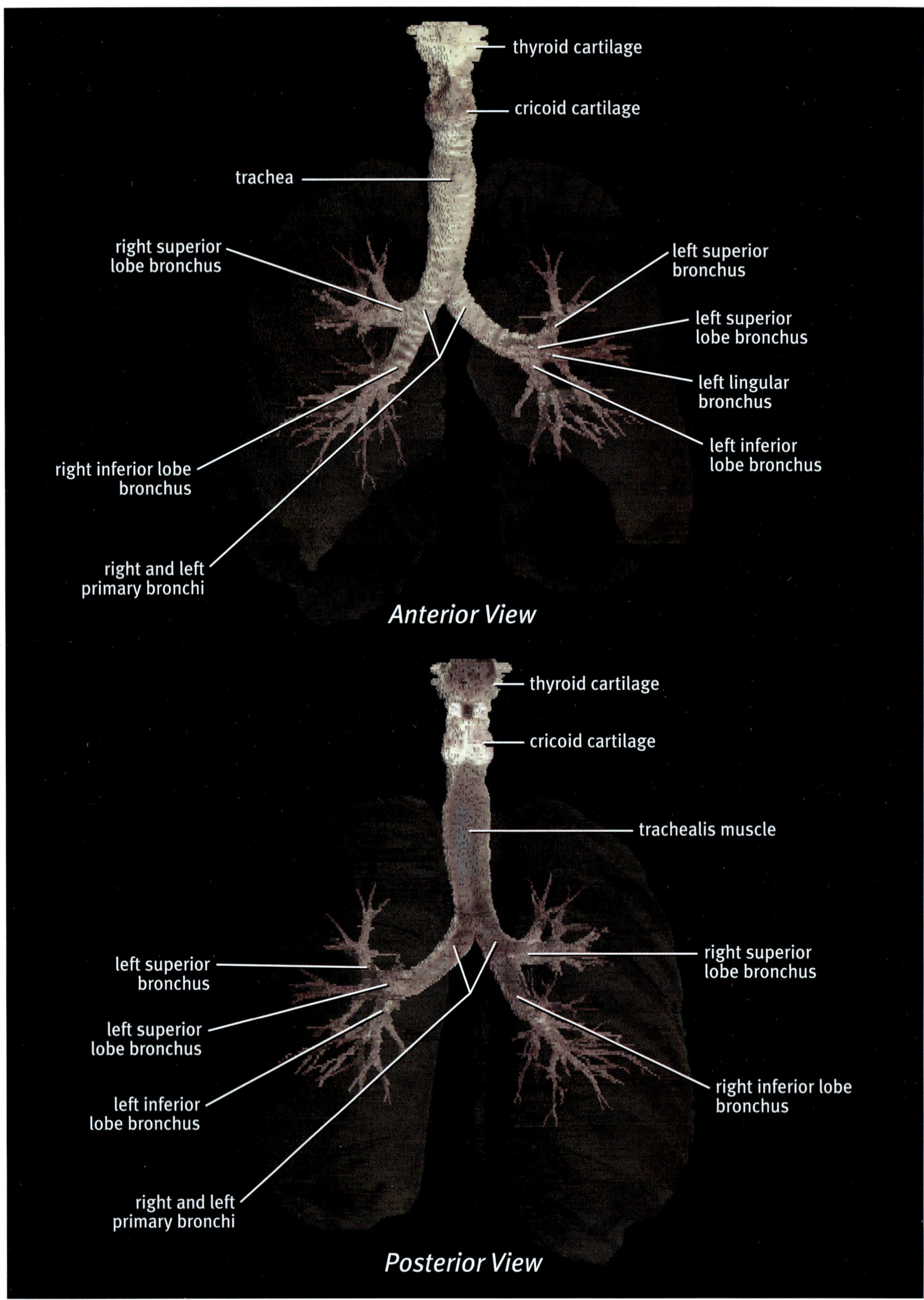

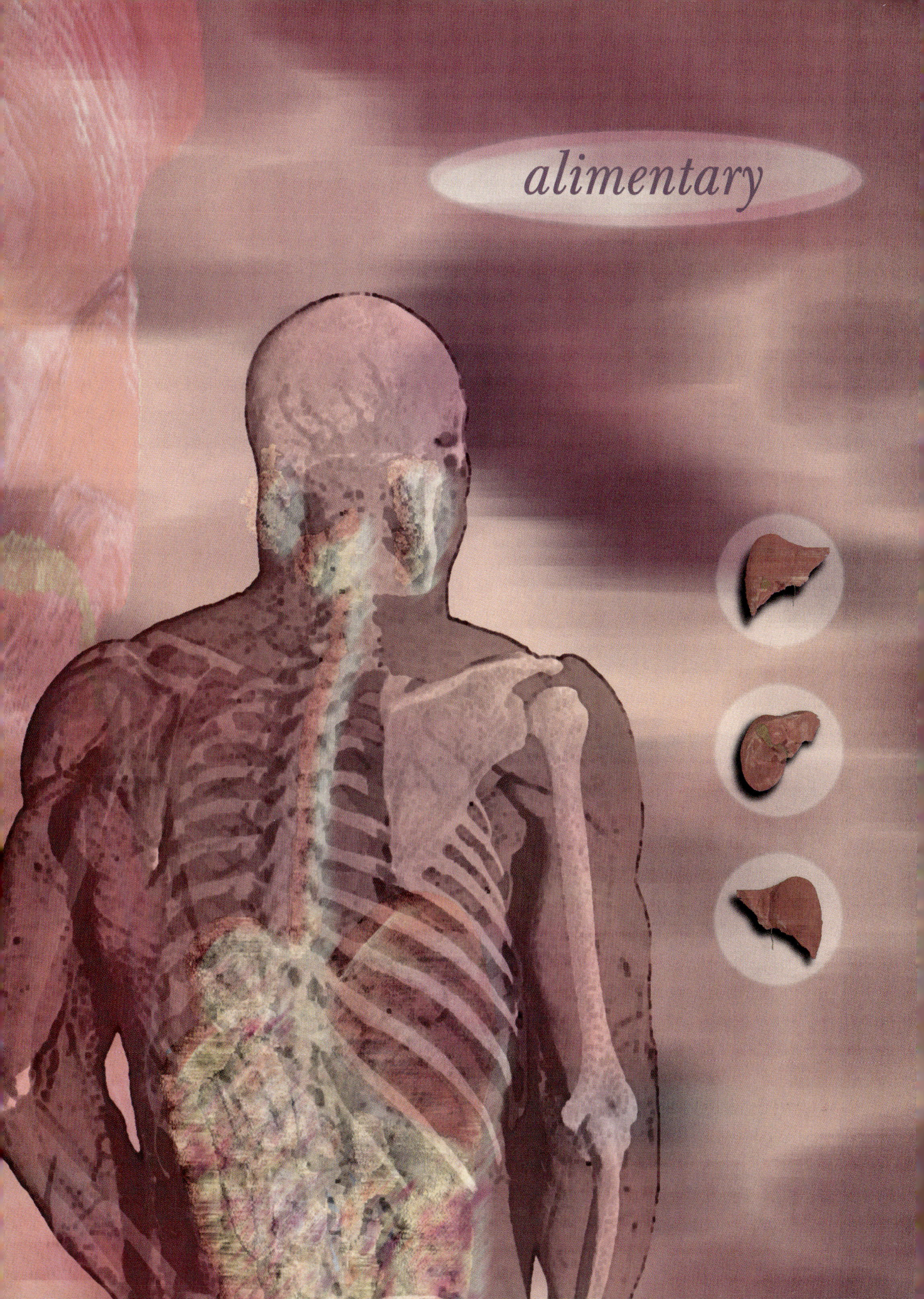
alimentary

Alimentary System 1

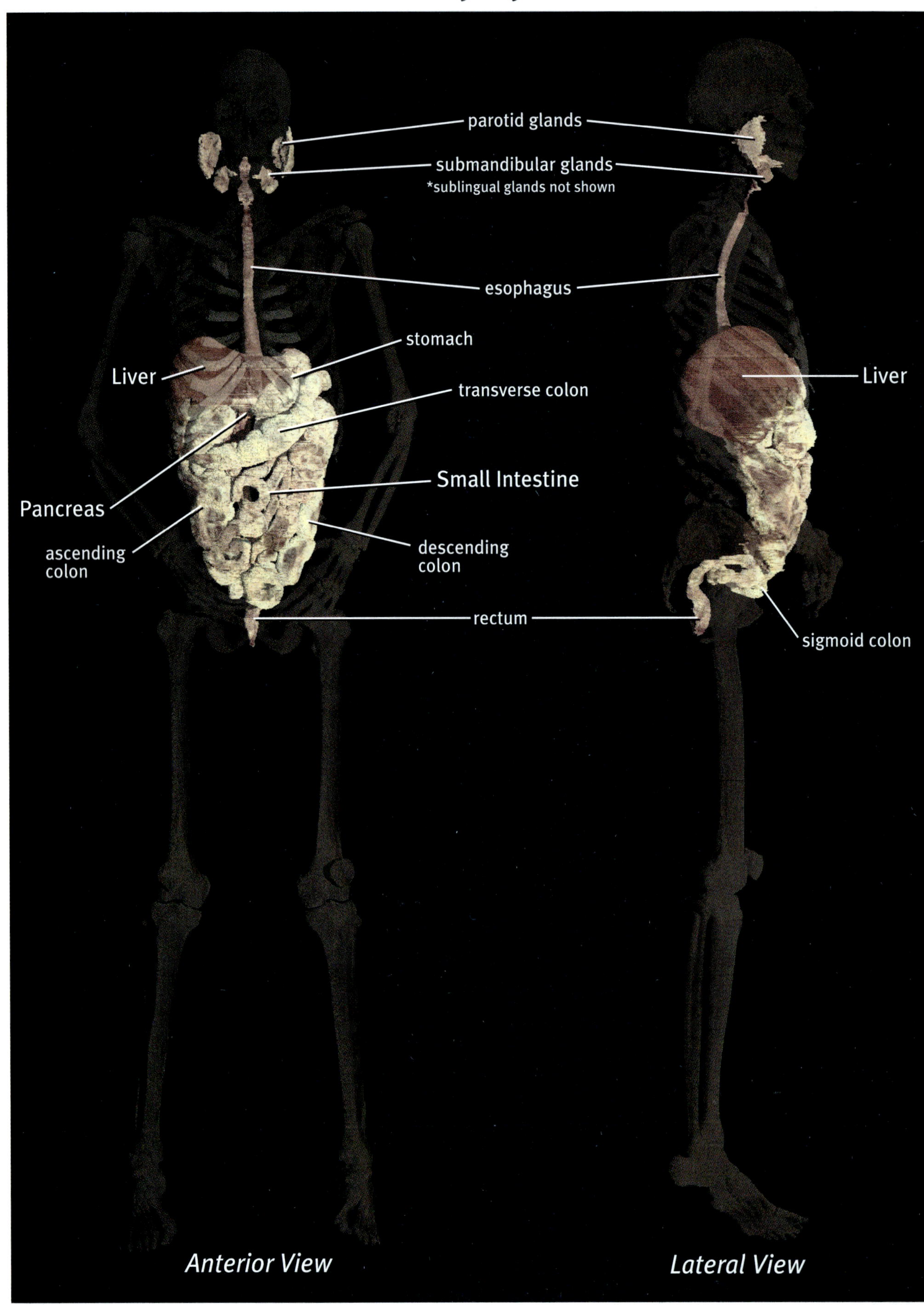

Alimentary System 2

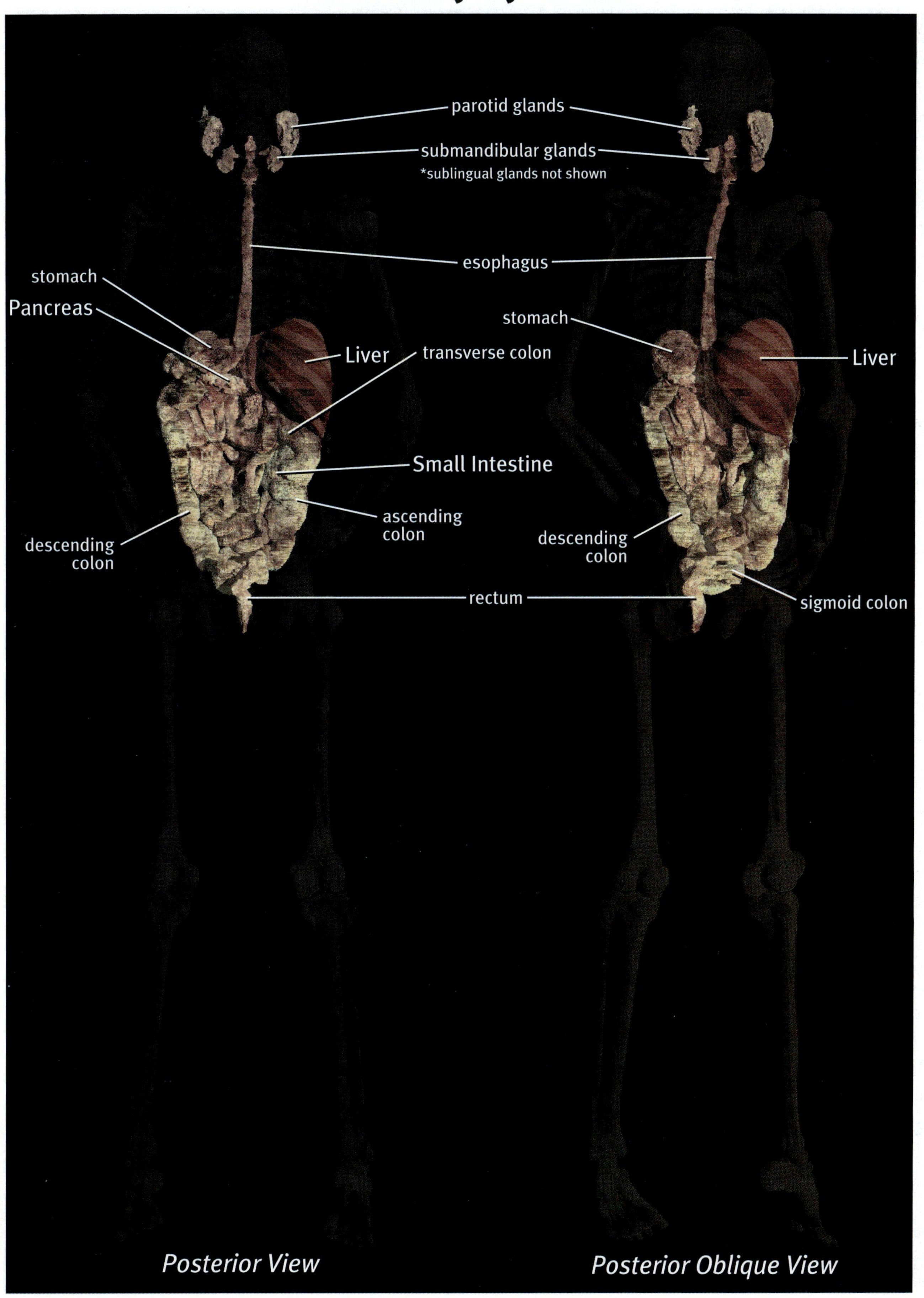

Upper Gastrointestinal Tract

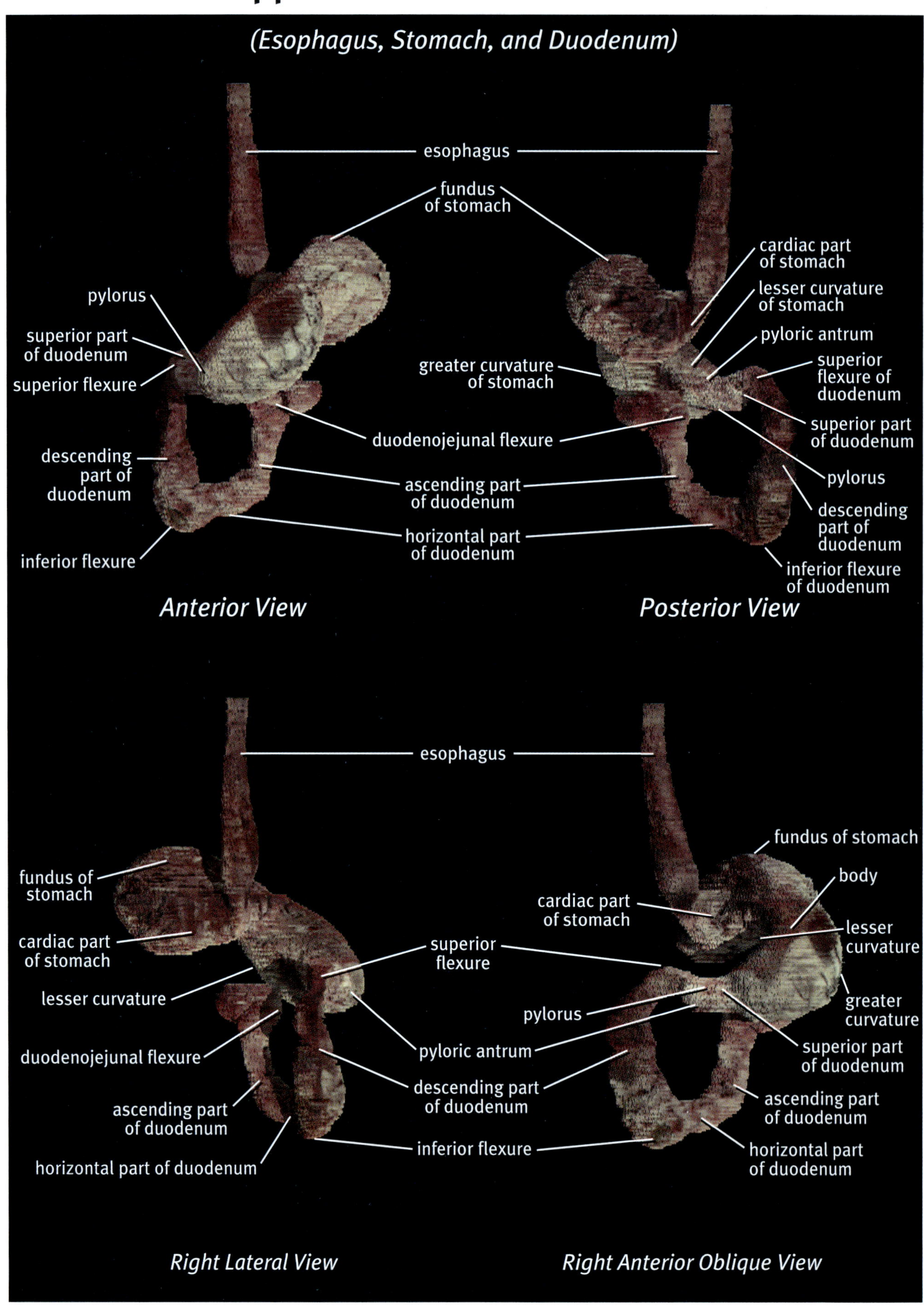

Small Intestine

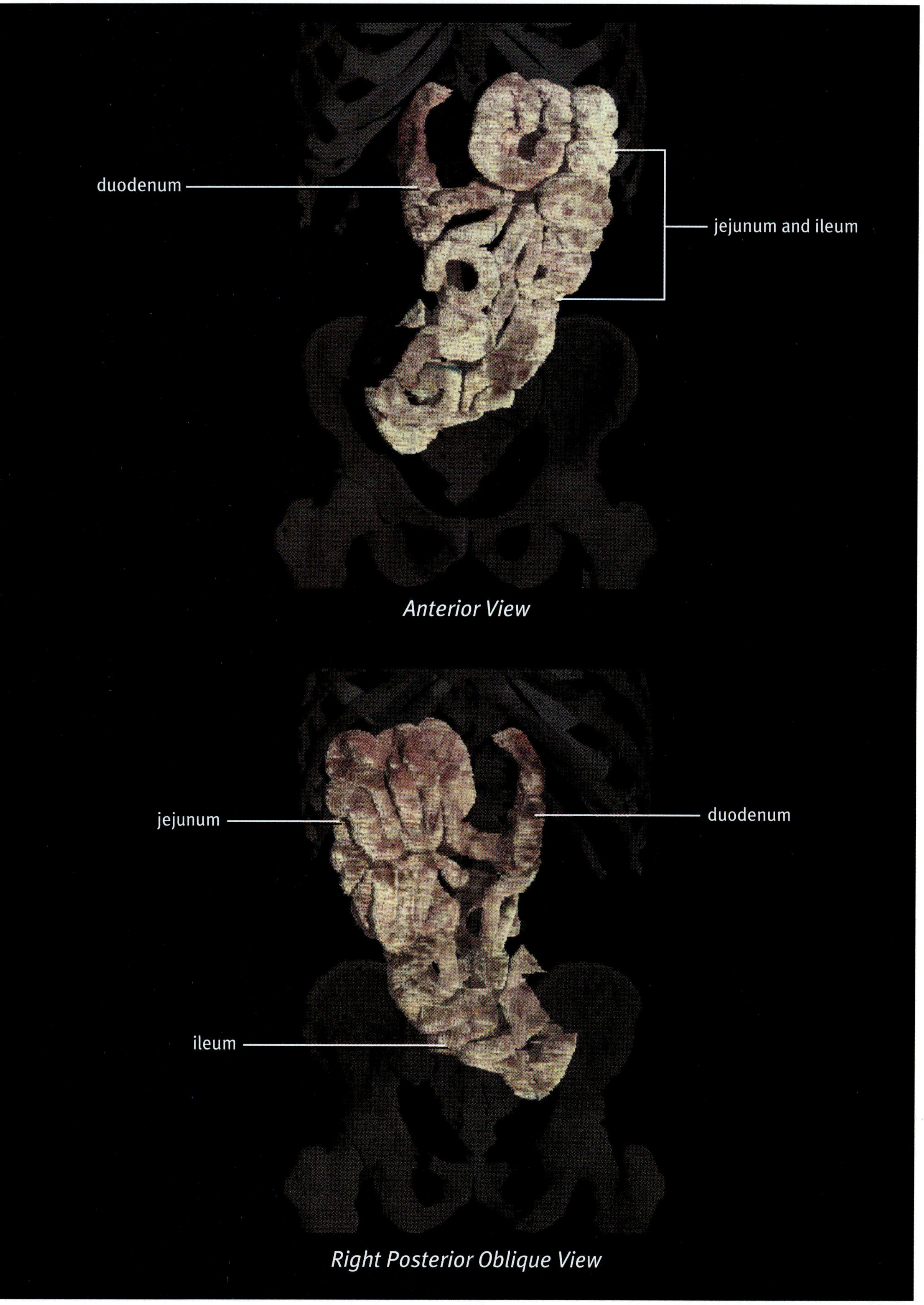

Large Intestine

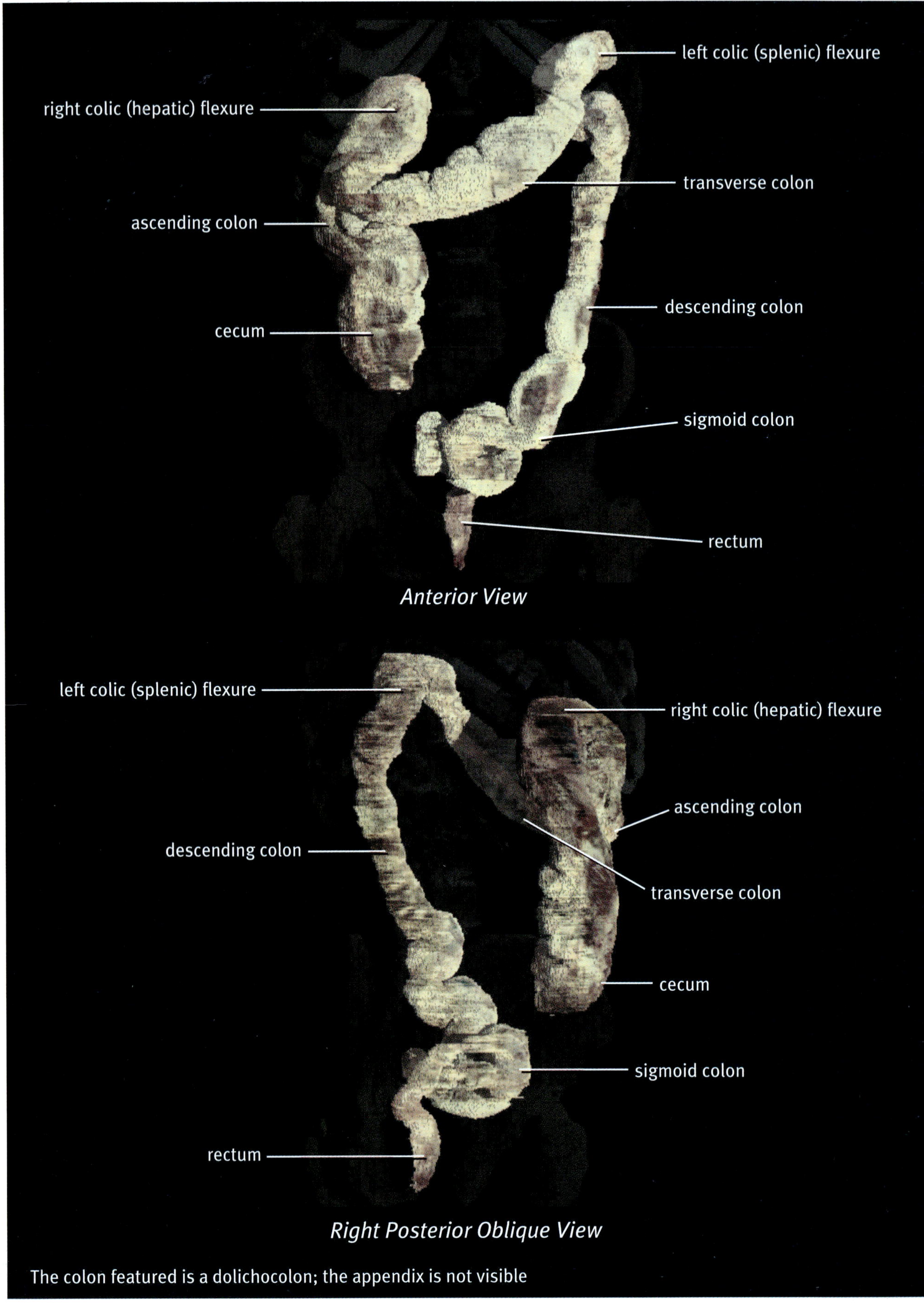

Anterior View

Right Posterior Oblique View

The colon featured is a dolichocolon; the appendix is not visible

Pancreas

body
tail
splenic vein
right
gastroepiploic
vein
neck
middle colic vein
head
Anterior View
branches of
splenic
vein
body
tail
splenic vein
head
Left Lateral View
tail
body
splenic vein
neck
right
gastroepiploic
vein
head
middle colic vein
Posterior View
Right Lateral View

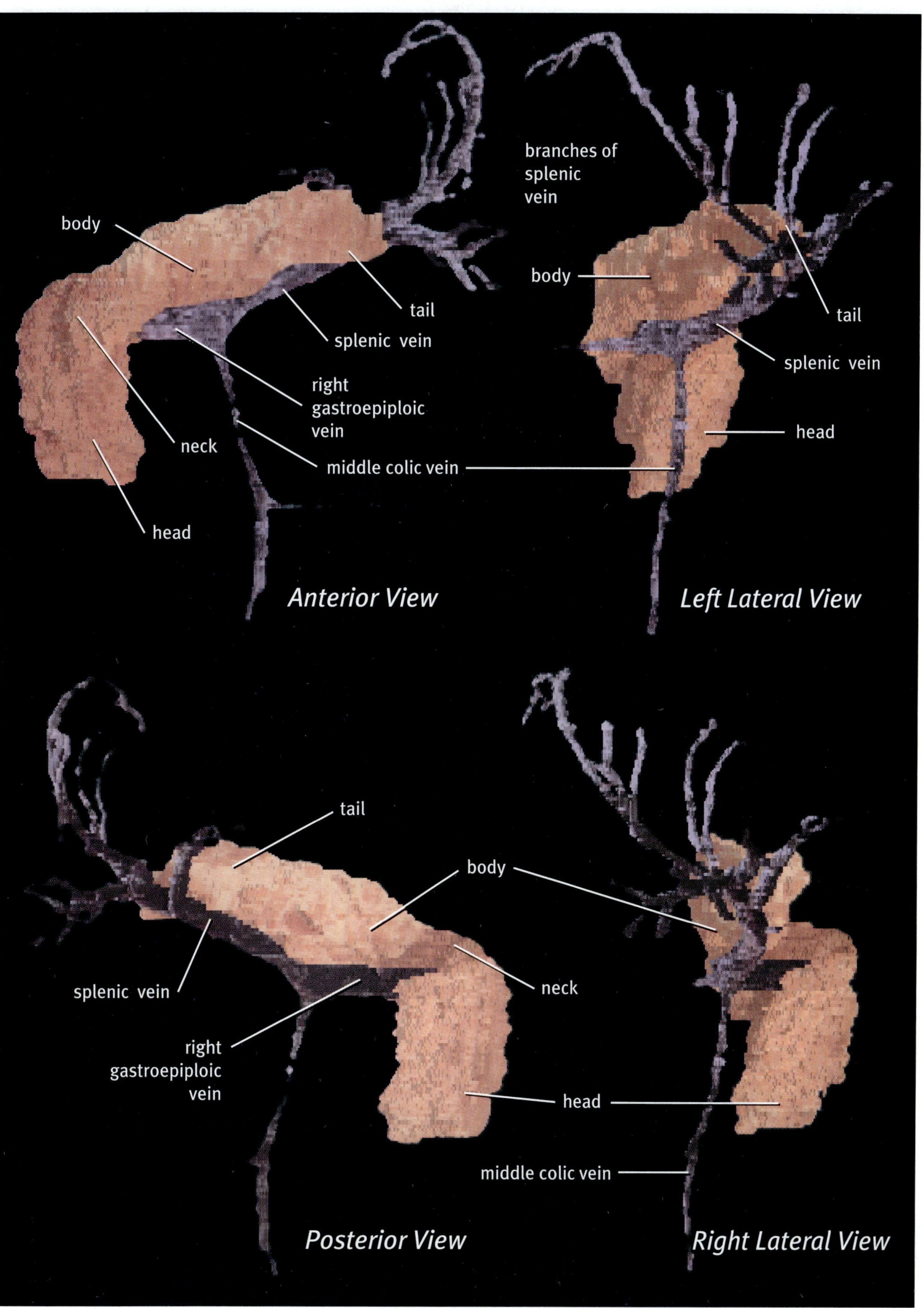

Liver 1

right lobe
left lobe
inferior margin of liver
costal impressions
gallbladder
common bile duct
inferior margin of liver
Anterior View
right lobe
left lobe
inferior margin of liver
common bile duct
inferior margin of liver
Posterior View

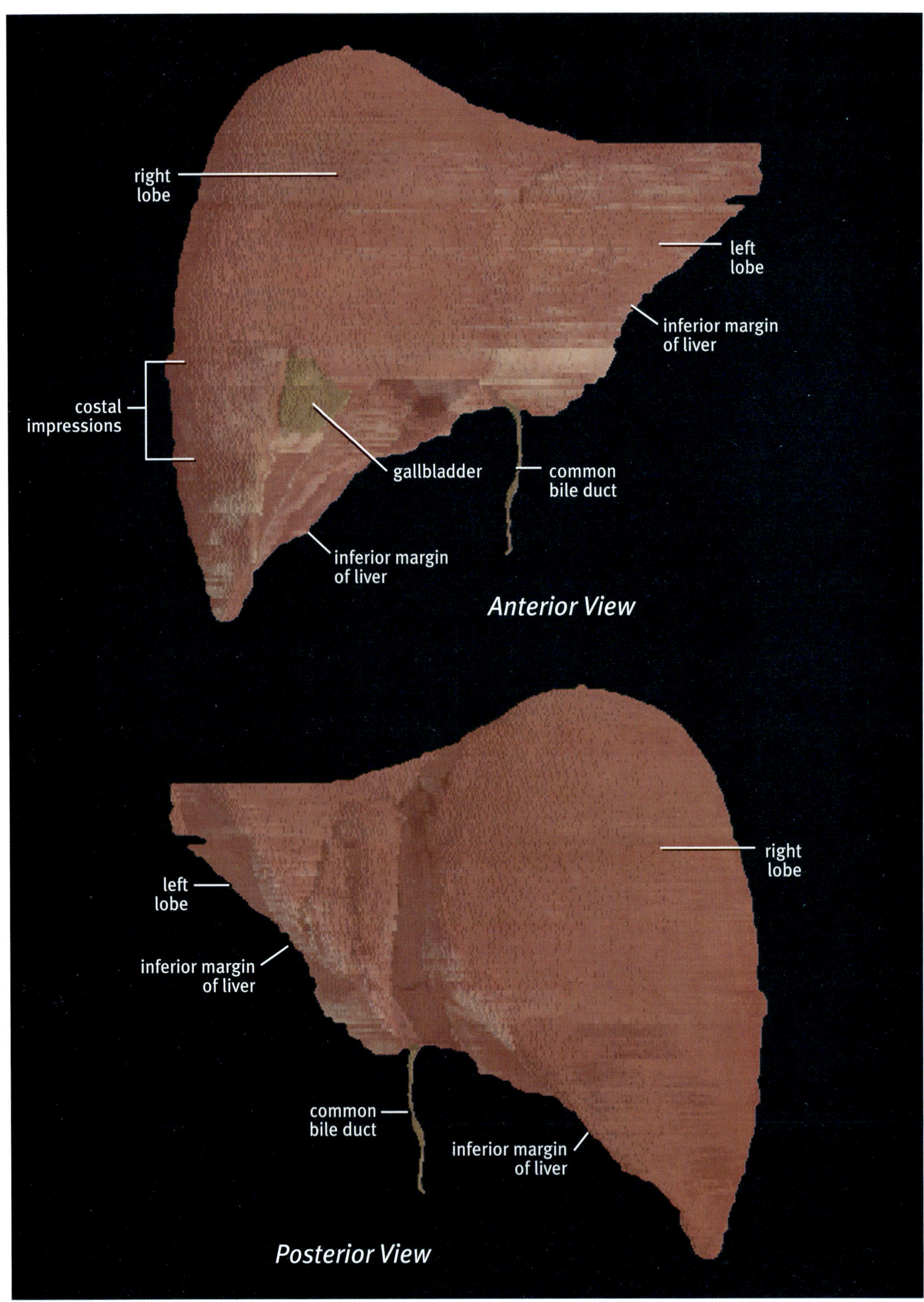

Liver 2

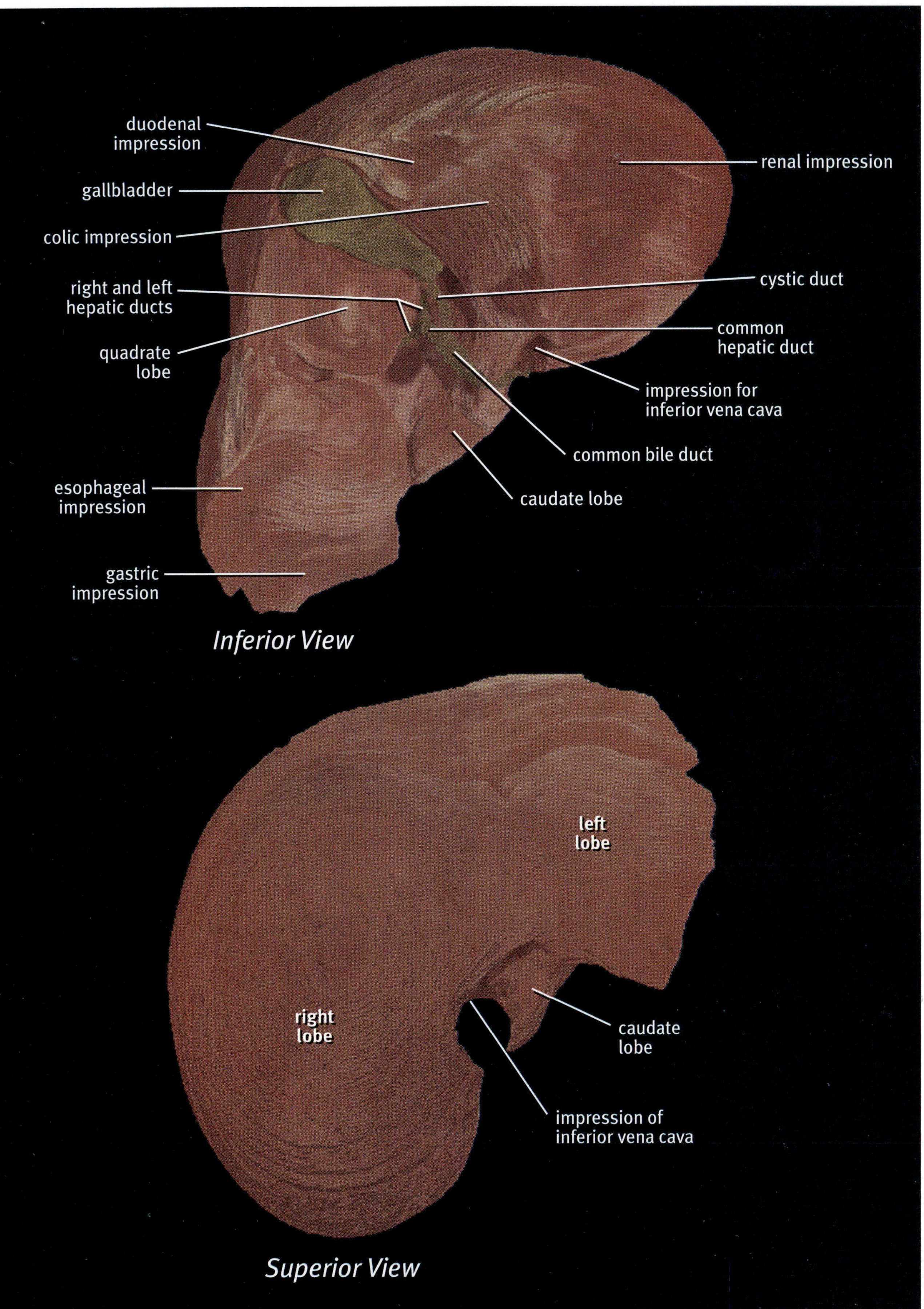

Inferior View

Superior View

Cross-sections of the Abdomen 1

Axial Section

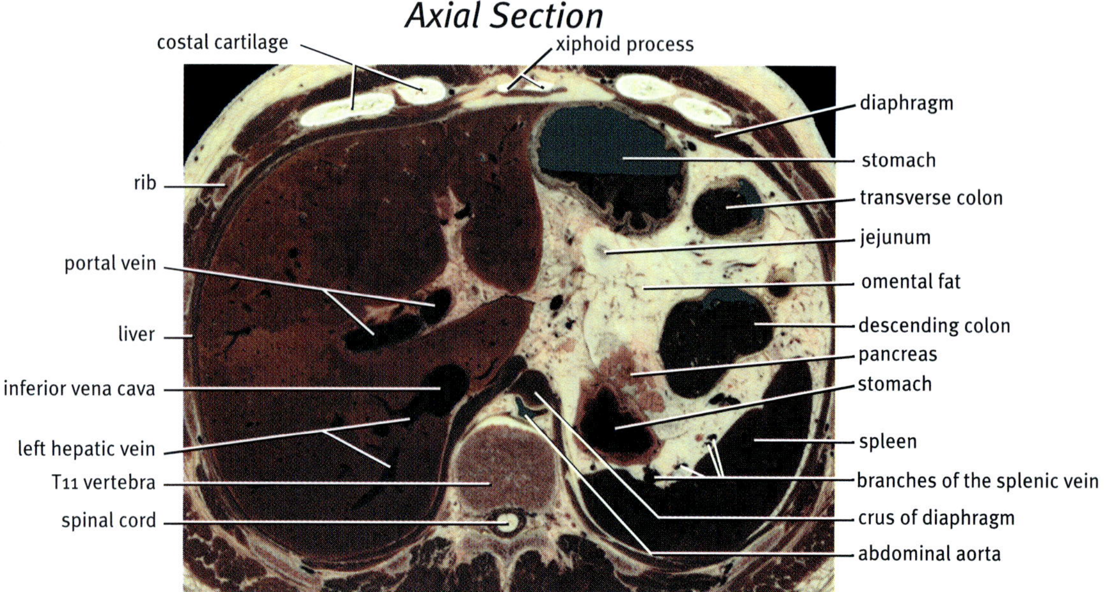

Coronal Section

right atrium
right lung
liver
gallbladder
pancreas
splenic vein
portal vein
ascending colon
external oblique muscle
internal oblique muscle
transverse muscle
obturator internus muscle
right ventricle
left lung
left ventricle
stomach
diaphragm
rib
small intestine (jejunum)
duodenum
small intestine (jejunum)
small intestine (ileum)

Cross-sections of the Abdomen 2

Axial Section

abdominal aorta
jejunum
ascending colon
descending colon
transverse muscle
inferior vena cava
internal oblique muscle
external oblique muscle
psoas muscle
L4 vertebra
multifidus muscle
spinous process of L3
quadratus lumborum muscle
iliocostalis muscle
longissimus dorsi muscle

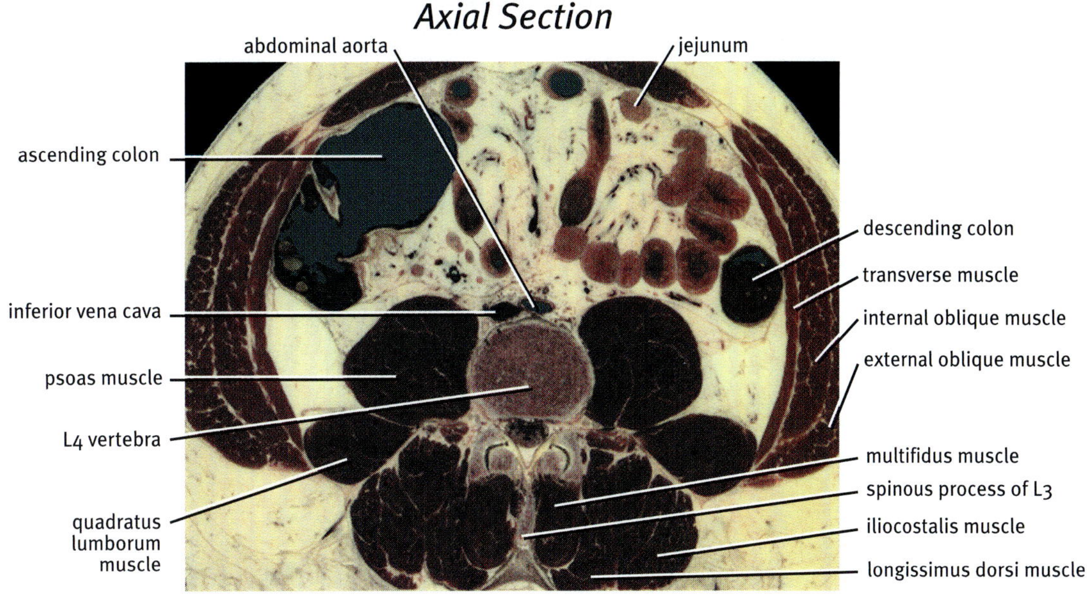

Sagittal Section

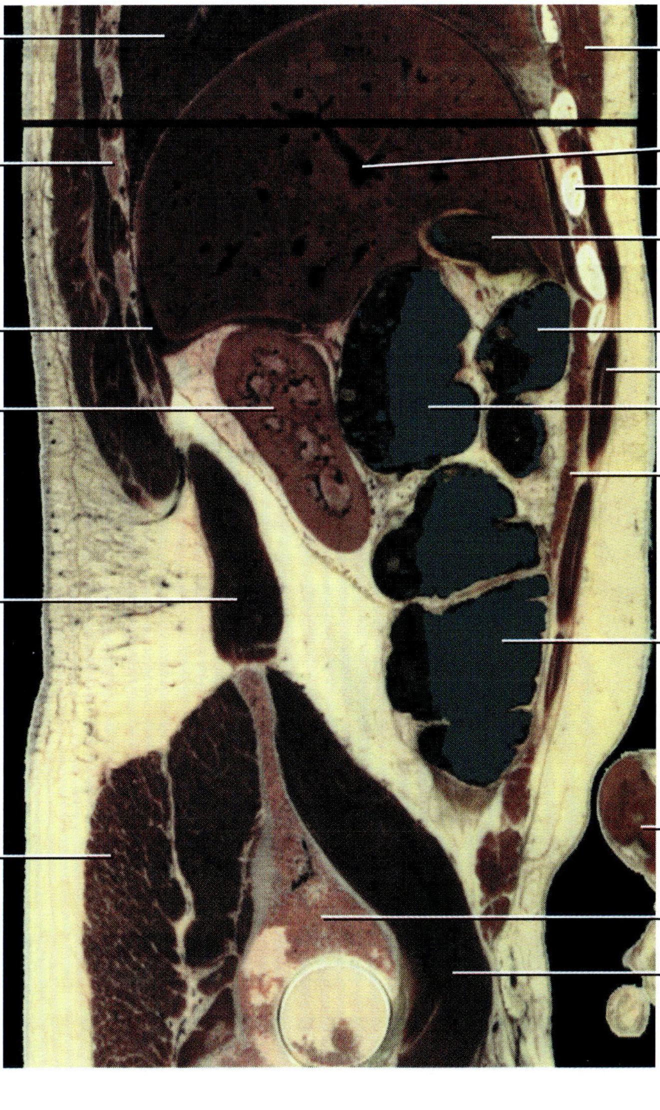

right lung
pectoralis major muscle
rib
branches of the portal vein
costal cartilage
gall bladder
capsule of Glisson
transverse colon
rectus abdominis muscle
right kidney
right (hepatic) flexure of the colon
transversus abdominis muscle
quadratus lumborum muscle
ascending colon (note haustrae)
hand
gluteus maximus muscle
pelvic bone
obturator internus muscle

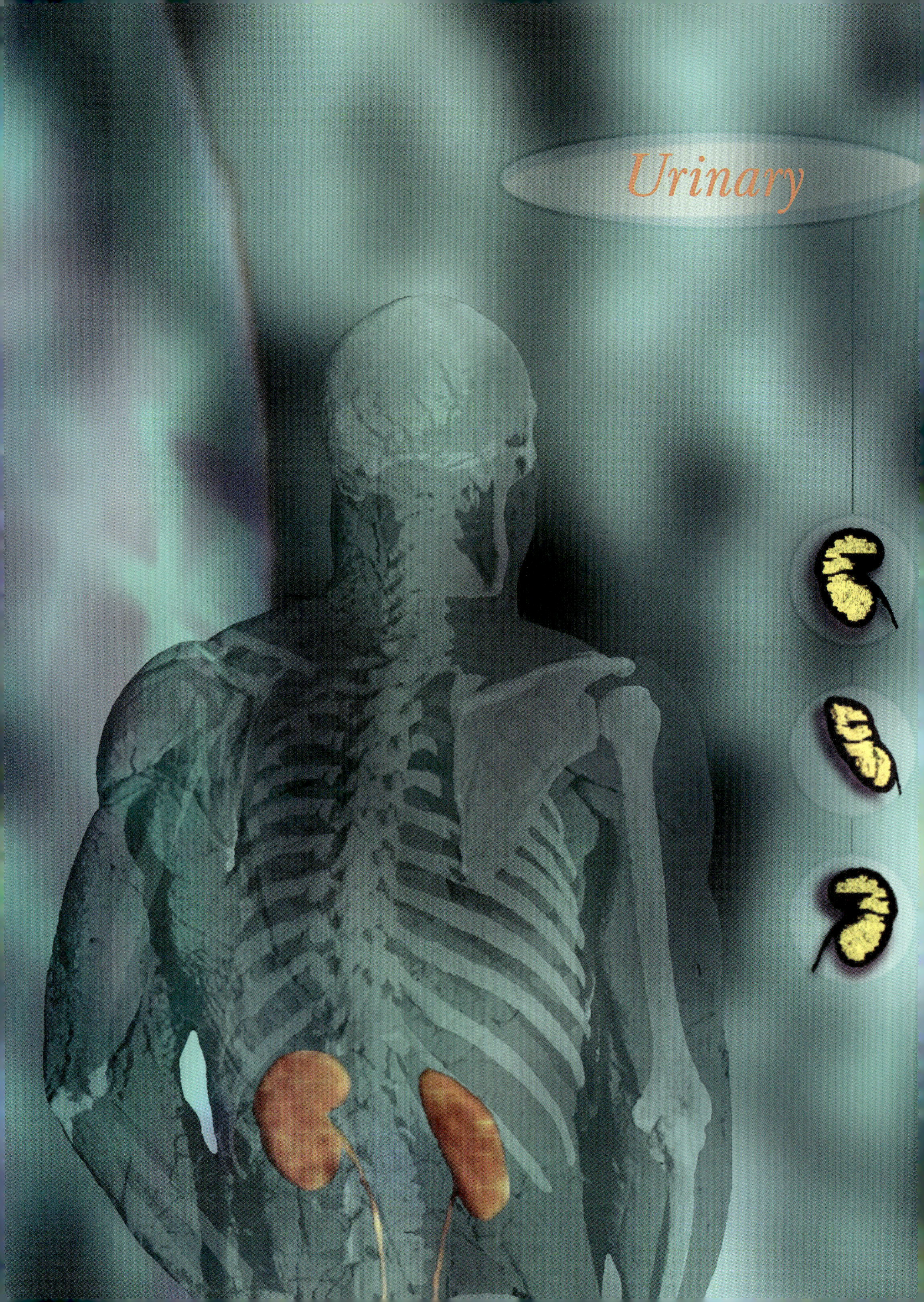
Urinary

Urinary System 1

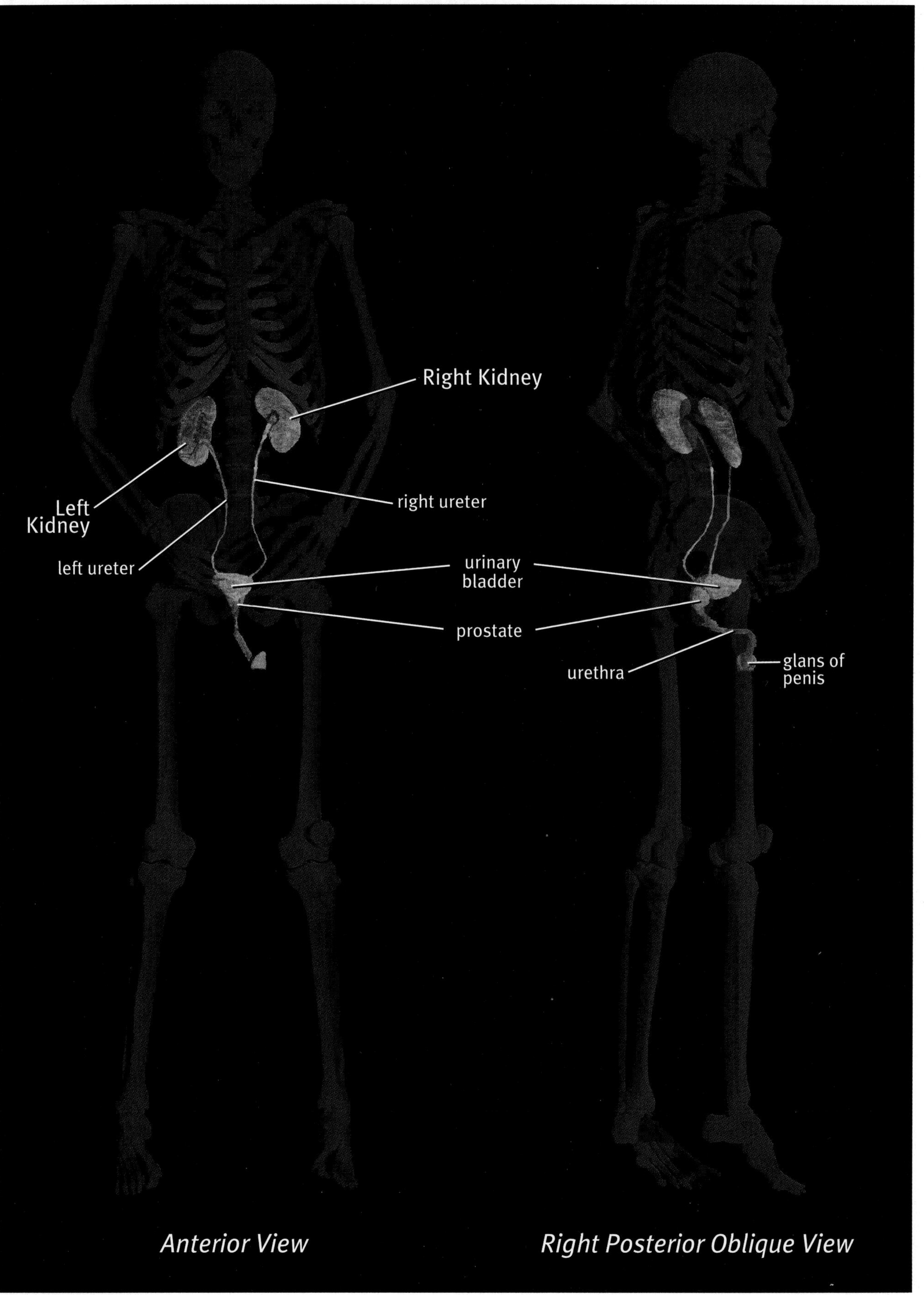

Anterior View

Right Posterior Oblique View

Urinary System 2

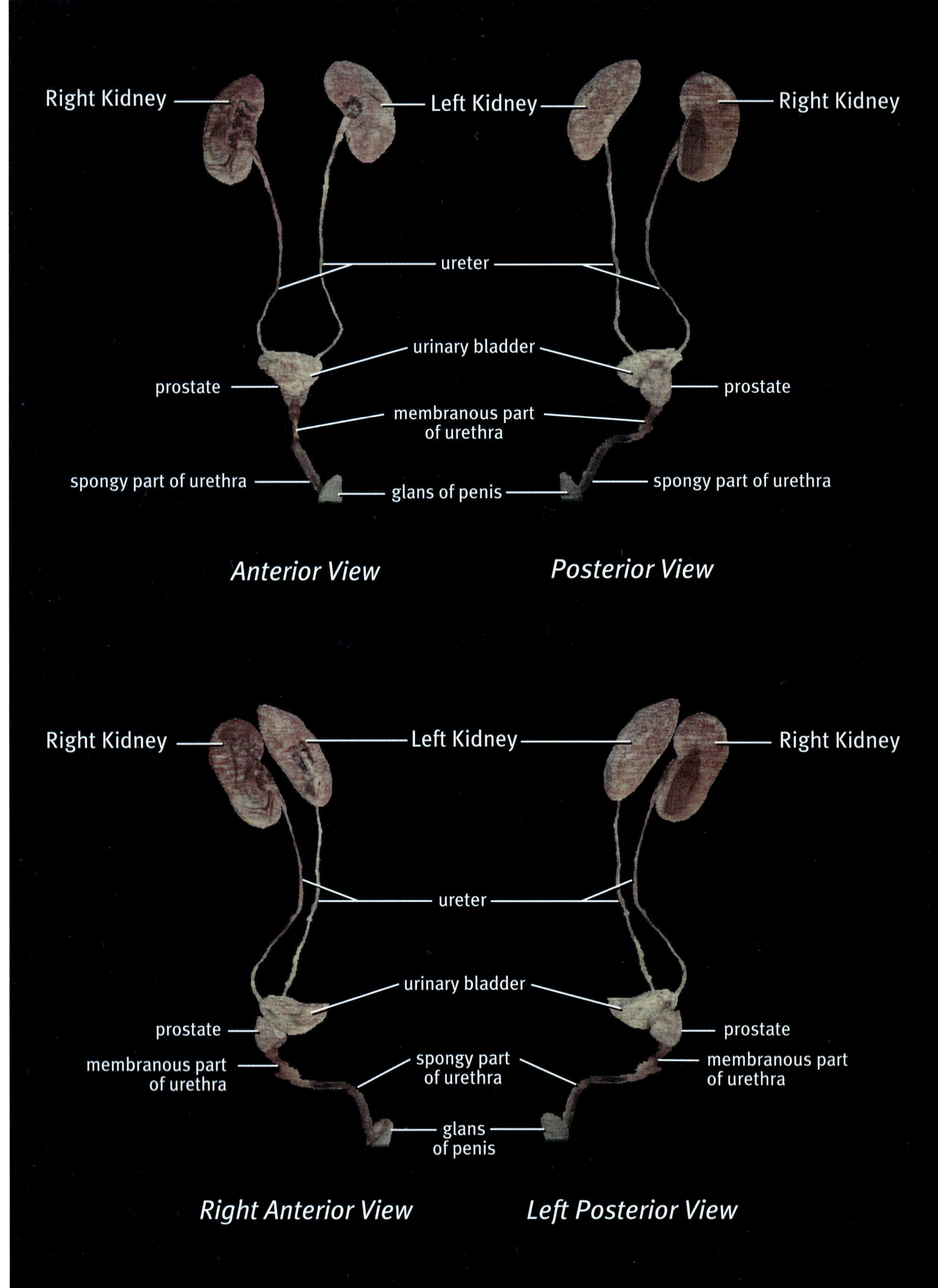

Cross-sections of the Kidney

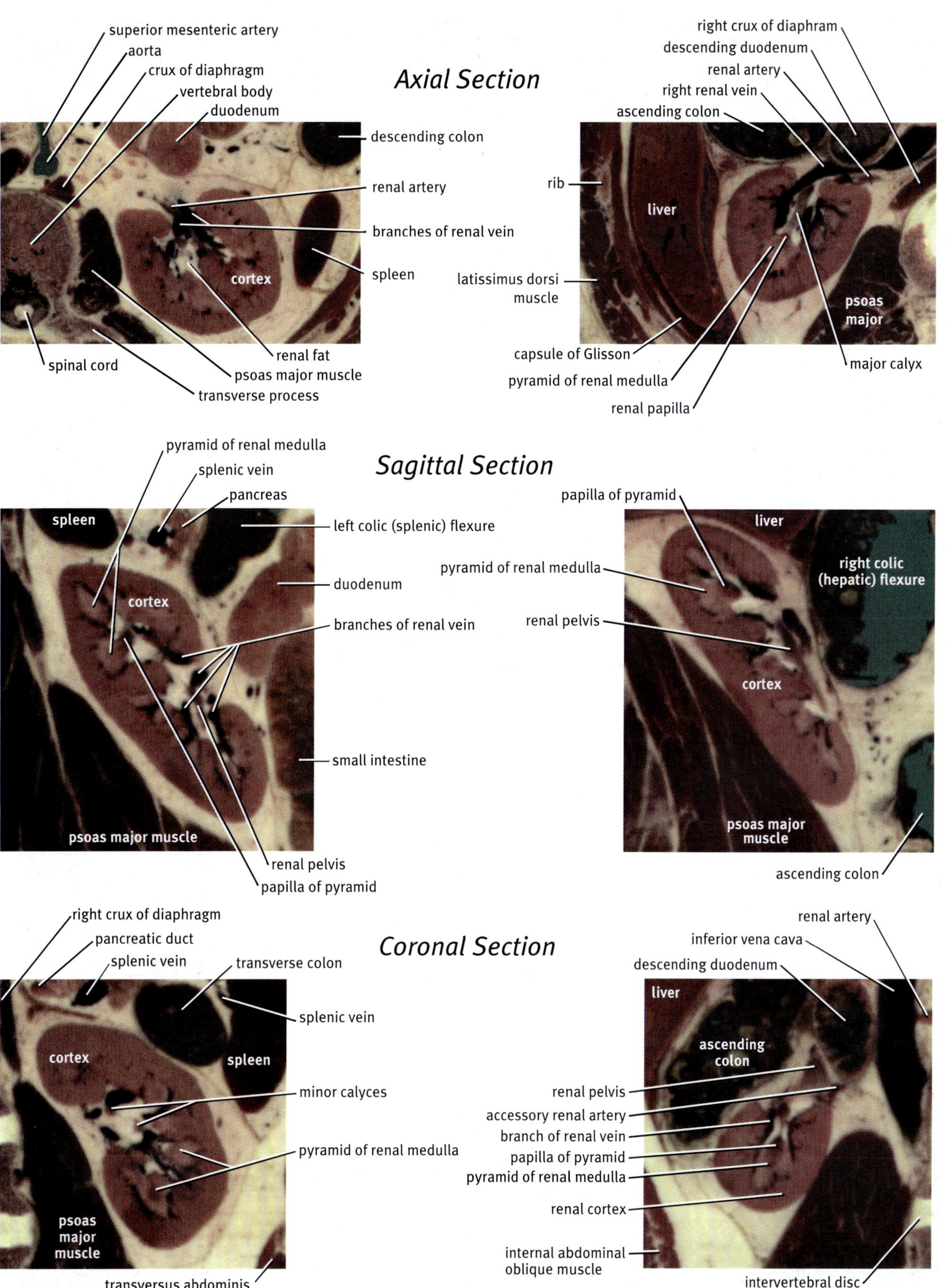

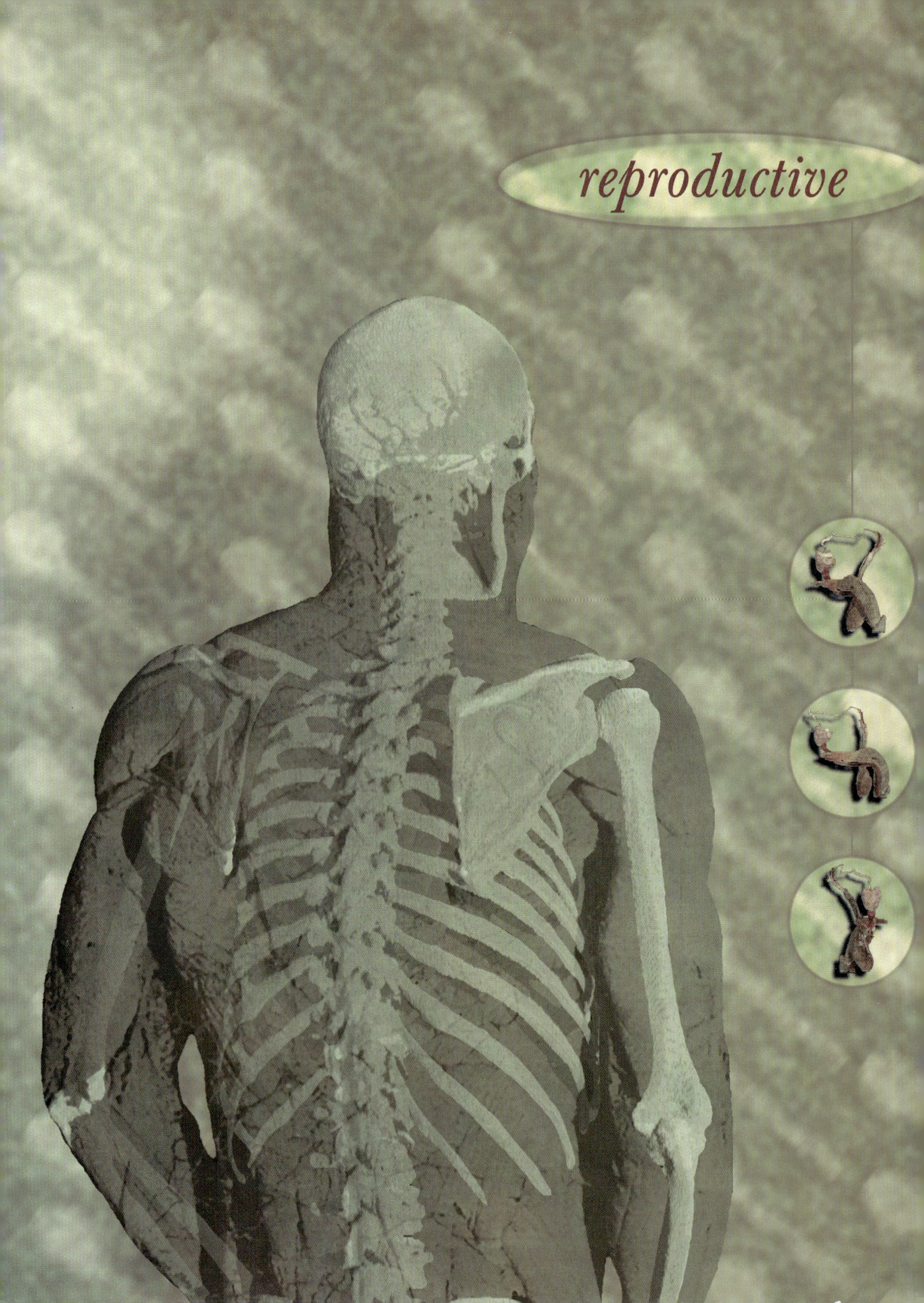

reproductive

Male Reproductive System

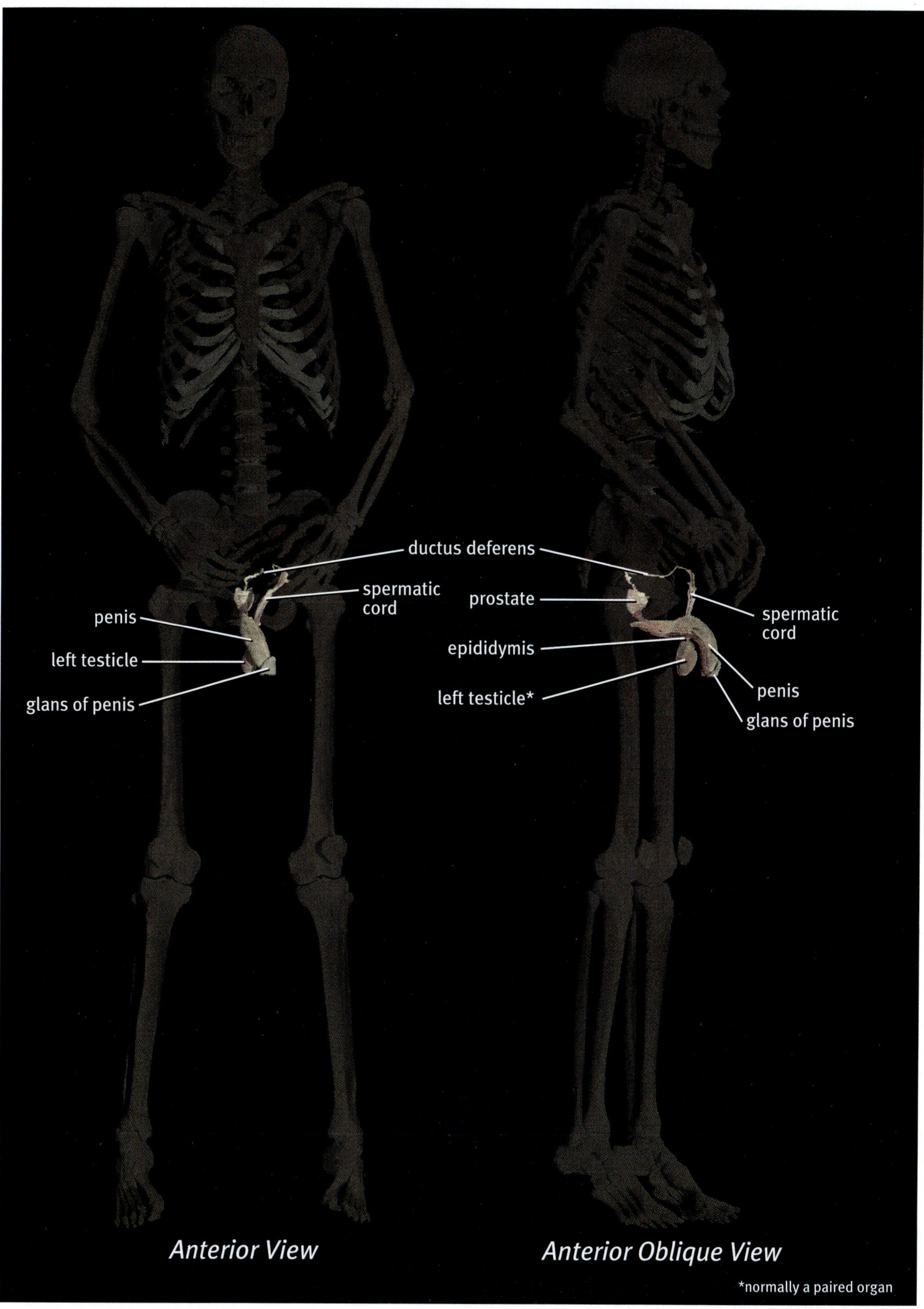

Male Genital Organs

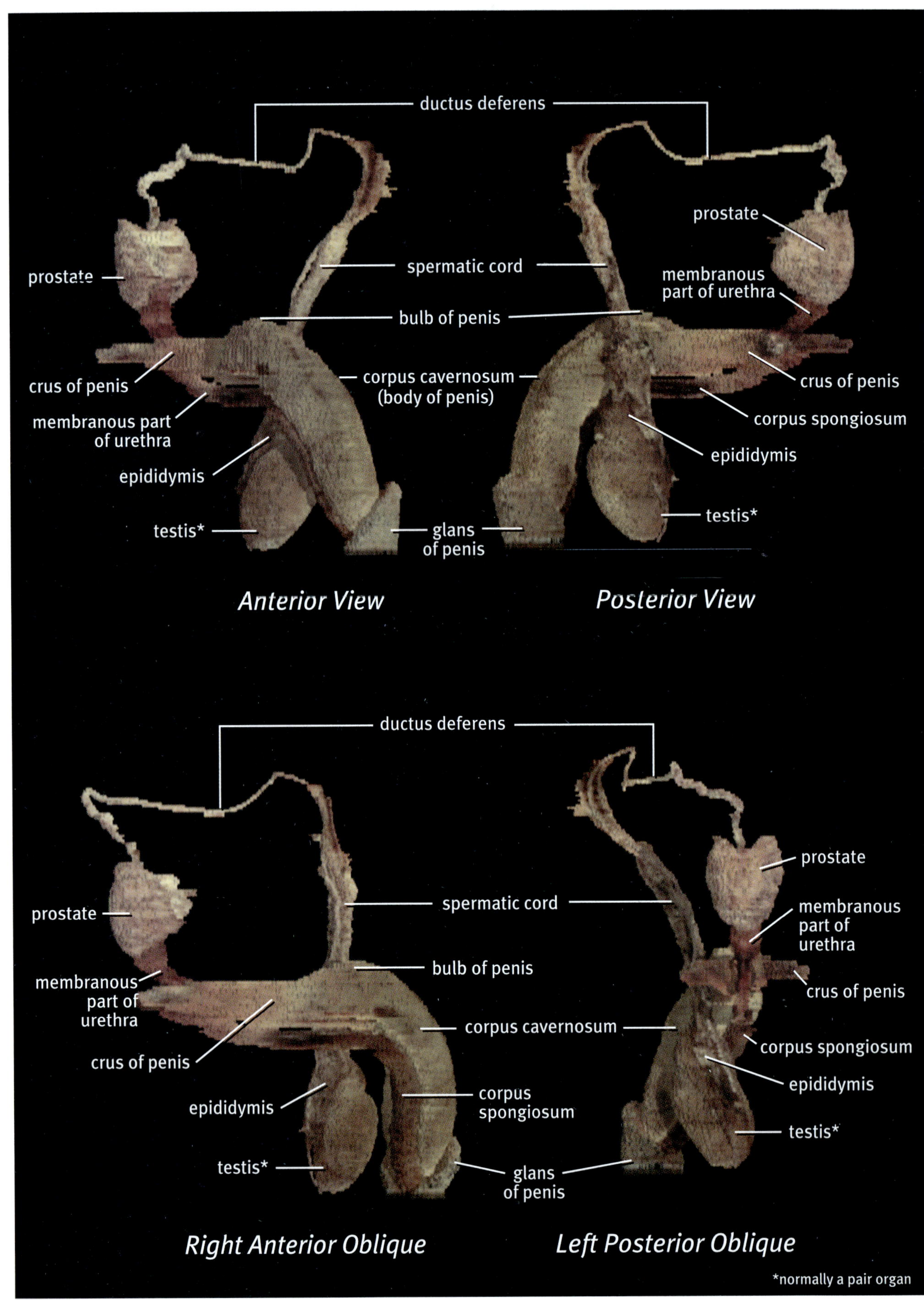

Cross-sections of the Male Pelvic Organs

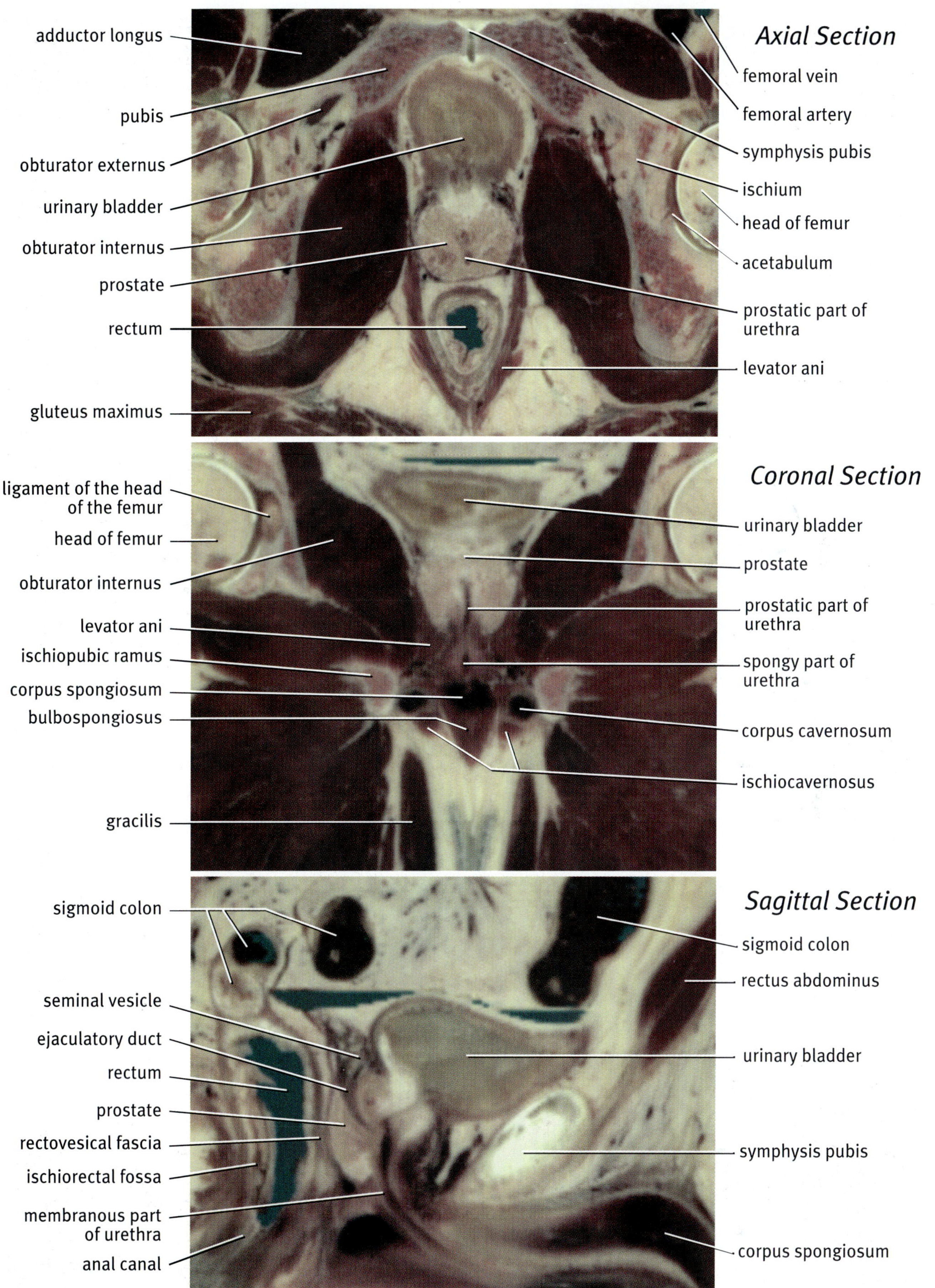

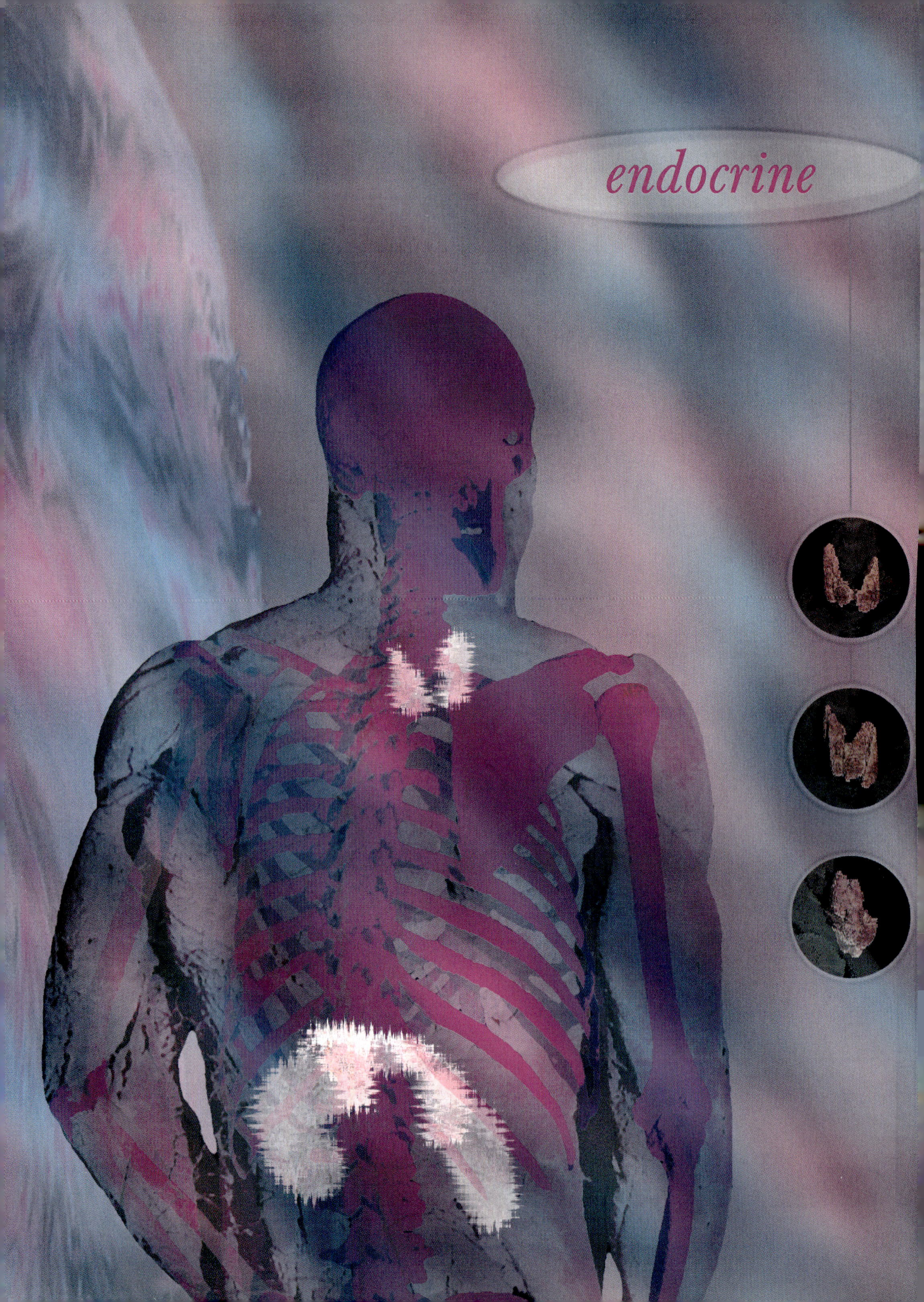
endocrine

Endocrine System

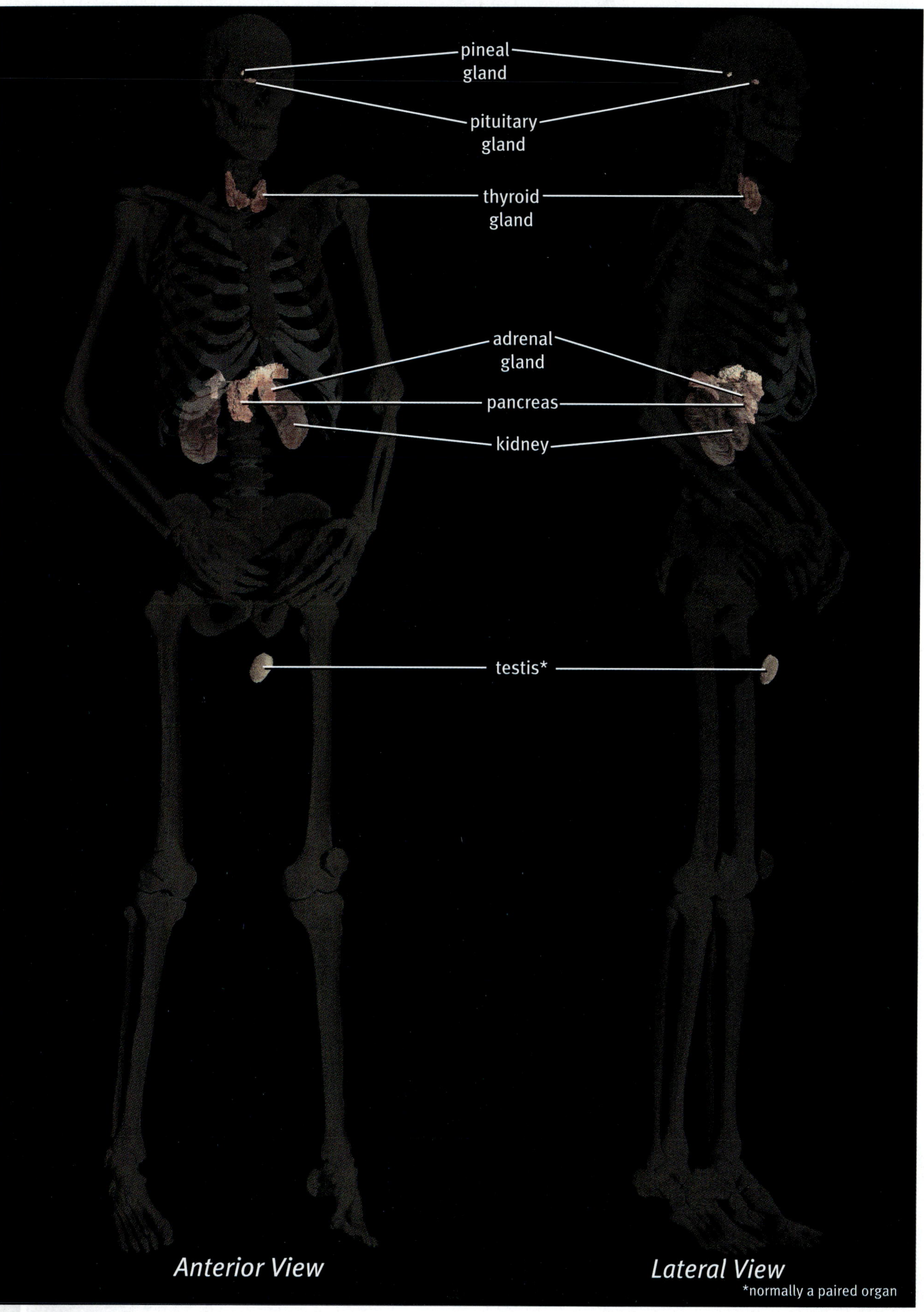

Anterior View

Lateral View

*normally a paired organ

Adrenal Glands

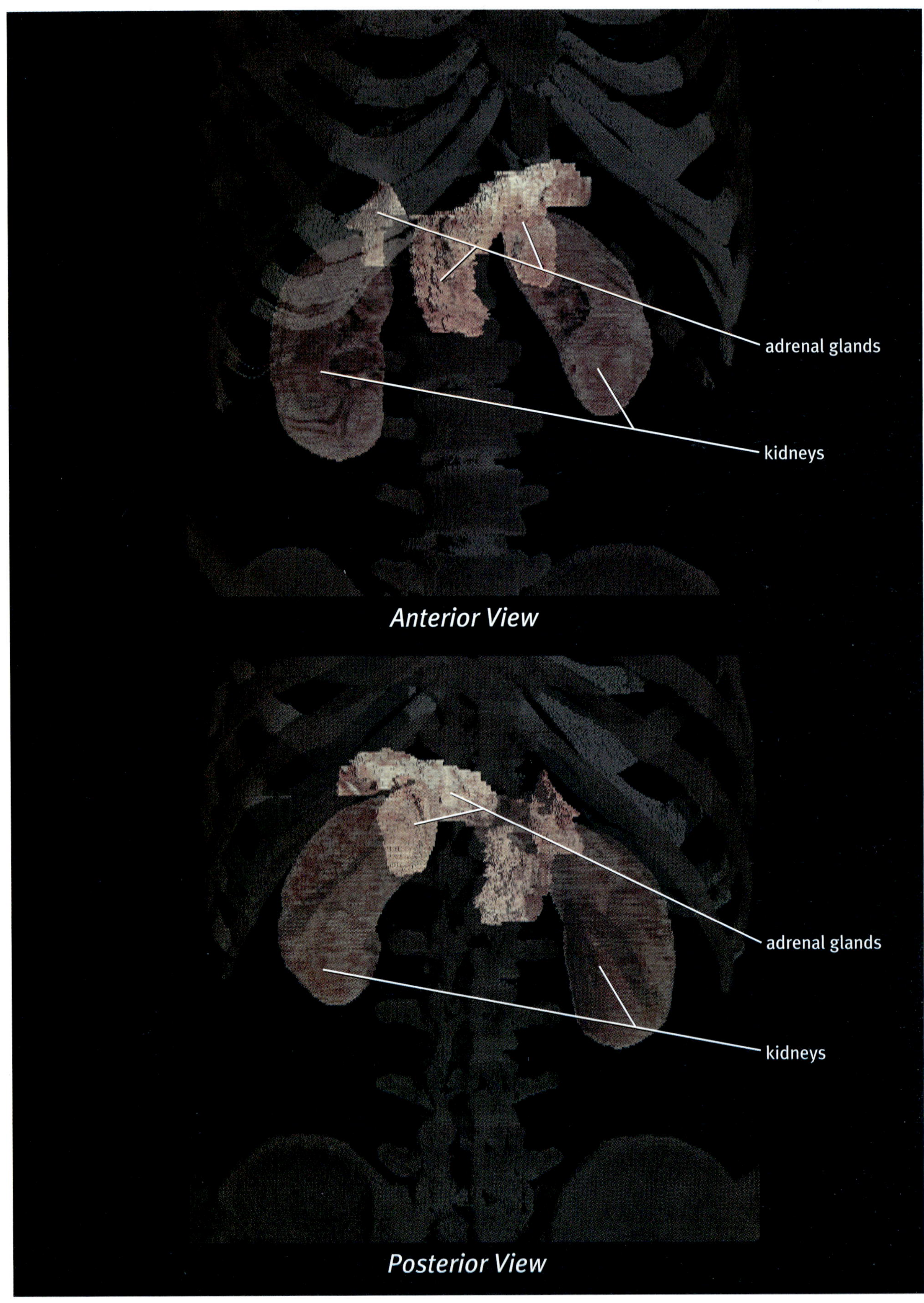

Anterior View

Posterior View

INDEX

bdomen 34
acromioclavicular joint 20
acoustic meatus, external 4, 6
acoustic meatus, internal 5
adrenal gland 117, 118
anterior clinoid process 5
anterior cranial fossa 5
anterior nasal spine 4
anterior tibial artery 73, 74
anterior tibial vein 73, 74
aorta 73, 74, 75, 76, 77, 78, 79, 82, 83, 107, 111
  abdominal 107
  arch 73, 74, 75, 82
  ascending 78, 82
  descending 73, 74, 75, 79, 82, 83
axilla 38
axillary artery 82
azygos vein 79, 83
back 34, 35
brachial artery 82
brachial plexus 60, 61, 70
brachial vein 82
brachiocephalic trunk 74, 75, 80. 81, 82
brachiocephalic vein 73, 82
brain, longitudinal fissure 62, 63
brainstem 60, 61, 62, 63, 64, 69
bronchus 96
  inferior lobe 96
  lingular 96
  secondary bronchus 79
  superior lobe 96
calcaneus 32, 58
capitate 24
carotid artery 73, 74, 75, 80, 81, 82
  common 73, 74, 75, 80, 81, 82
  external 80
  internal 80
arotid canal 6
audate nucleus 67, 68
ecum 102
entral sulcus 64
ephalic vein 46, 73, 74, 80
erebellar peduncle 69
erebellum 60, 61, 62, 63, 64, 67, 68, 69
erebral aqueduct 65, 67
erebral hemisphere 63, 64
erebral peduncle 67
erebrum 60, 61, 63
est 34
oanae 6
gulate gyrus 64, 68, 69
vicle 16, 18, 19, 20, 21, 46
  acromial end 21
  conoid tubercle 21
on 98, 99, 102, 107, 111
  ascending 98, 99, 102, 107, 111
  descending 98, 99, 102, 107, 111
  hepatic (right) flexure 102, 107, 111
  sigmoid 98, 99, 102
  splenic (left) flexure 102, 111
  transverse 98, 99, 102, 107, 111
mmon bile duct 104, 105
mmon hepatic duct 105
dylar canal and fossa 6
ona radiata 68
onary sinus 75, 78, 79
onary sulcus 75
onary vein 79
ous callosum 64, 67, 68, 69
al cartilage 15, 79, 107
oid cartilage 92, 93, 94, 95, 96
s sections 46, 51, 57, 58, 67-69, 78, 79, 106, 107, 111
  abdomen 106, 107
  ankle 58
  brain 67, 68, 69
  heart 78, 79
  hip 51
  kidney 111
  knee 57
  pelvic organs 115
  shoulder 46
iate ligament, anterior 57
iate ligament, posterior 57
oid 32
eiform 32, 58

  intermediate 32, 58
  lateral 32
  medial 32
cystic duct 105
deltoid tuberosity 22
deltopectoral sulcus 46
diencephalon 63
ductus deferens 113, 114
duodenojejunal flexure 100
duodenum 100, 101, 111
epididymis 113, 114
esophagus 78, 79, 98, 99, 100
ethmoid, perpendicular plate 6
external occipital crest 6
external occipital protuberance 4, 6
eyeball 62, 63, 64
facial vein 80
femoral artery 51, 73, 74
femoral vein 51, 73, 74
femur 25, 27, 28, 29, 51, 57
  adductor tubercle 28
  fovea of head 28, 29
  head 28, 29, 51
  intercondylar fossa 28, 29
  lateral condyle 28, 29, 57
  lateral epicondyle 28, 29
  lesser trochanter 28, 29
  medial condyle 28, 29, 57
  medial epicondyle 28, 29
  neck 28, 29
  patellar surface 28, 29
  shaft 28
  trochanteric fossa 28
fibula 25, 30, 31
  articular surfaces 30, 31
  head 30, 31
  interosseous border 31
  lateral malleolus 30, 31, 58
  malleolus 31
  medial crest 30
  neck 30, 31
  shaft 30
foot, phalanx, phalanges 32
foramen 5, 6
  cecum 5
  incisive 6
  lacerum 5, 6
  magnum 5, 6
  ovale 5, 6
  spinosum 5
fornix 68
frontal bone 4
frontal bone, orbital surface 4
frontal lobe 62, 67, 68, 69
frontal sinus 5
Glisson, capsule of 107, 111
gallbladder 104, 105, 107
gastroepiploic vein (right) 103
glabella 4
glenoid 46
glenoid cavity 20
globus pallidus 67
gluteal tuberosity 28, 29
greater curvature 100
greater palatine foramina 6
greater saphenous vein 57, 58, 73, 74
greater sciatic notch 26, 27
greater trochanter 28, 29, 51
greater tubercle 22
hamate bone 24
hand 18, 19, 24, 107
  phalanx, phalanges 24
head and neck 34, 35
heart 73-79, 82
  anterior interventricular sulcus 75
  anterior interventricular artery 75
  atrioventricular valve 77
  atrium 75, 76, 77, 78, 79, 82
  auricle (right) 75, 77, 78
  bicuspid valve 78, 79
  left ventricle 73, 74, 75, 76, 77, 78, 79, 82
  papillary muscle 79
  right atrium 73, 74
  right ventricle 75, 76, 77, 78, 79, 82
  tricuspid valve 78
hepatic ducts 105
hepatic portal and venous system 83
hepatic vein 83, 87

hippocampus 67, 69
horizontal fissure 95
humerus 18, 19, 22, 46
  anatomical neck 22
  coronoid fossa 22
  head 22, 46
  lateral epicondyle 22
  lateral supracondylar ridge 22
  medial epicondyle 22
  medial supracondylar ridge 22
  neck 22
  olecranon fossa 22
  trochlea 22
hypoglossal canal 6
hypophysis (pituitary gland) 63
ileum 101
iliac artery 73, 74, 82, 83
  common 73, 74, 82, 83
  external 73, 74
  internal 73
iliac crest 26, 27, 38, 52, 53, 54
iliac spine, posterior superior 26, 27
iliac vein 73, 74, 82, 83
  common 74, 82, 83
  external 73
  internal 73, 74
iliofemoral ligament 51
iliopubic eminence 26, 27
ilium 26, 27, 51
  anterior iliac spine 27
  articular surface 27
  body 27
  wing 26, 27
inferior articular surface 30, 31
inferior orbital fissure 4
infraorbital foramen 4
infrapatellar fat pad 57
innominate bone 25, 26, 27, 51, 107
  acetabular margin 27
  acetabular notch 27
  acetabulum 26, 27, 51
  lesser sciatic notch 27
  obturator crest 27
  obturator foramen 26, 27
  symphyseal surface 27
interatrial septum 78
internal capsule 67, 68, 69
interosseus talocalcanear ligament 58
interthalamic adhesion 65
interventricular foramen 65
interventricular sulcus, posterior 75
intervertebral disc 111
ischial spine 26, 27
ischiatic tuberosity 26, 27
ischiofemoral ligament 51
ischium 27, 51
jejunum 101, 107
jugular foramen 5, 6
jugular fossa 6
jugular vein 73, 80
  external 73, 80
  common trunk (right) *atypical* 80
  internal 80
kidney 83, 107, 109-111, 117, 118
  hilus 110
  major calyx 111
  minor calyces 111
  papilla of pyramid 111
knee, joint capsule 57
lateral collateral ligament 57
lateral talocalcanear ligament 58
lesser saphenous vein 58
ligament of head of femur 51
liver 98, 99, 104, 105, 111
  colic impression 105
  duodenal impression 105
  gastric impression 105
  impression of inferior vena cava 105
  renal impression 105
lower appendicular skeleton 2, 3
lower extremity 3, 34, 35
lumbosacral plexus 60, 61, 71
lunate 24
lung 46, 78, 79, 92, 94, 95, 107
  apex 92, 94, 95
  cardiac impression 95
  diaphragmatic surface 94, 95
  impression for esophagus 95, 105

  impression for trachea 95
  inferior lobe (left) 92, 93, 94, 95
  inferior lobe (right) 92, 94, 95
  left 46, 78, 79, 106
  lingula 94, 95
  oblique fissure 94, 95
  right 79, 93, 106, 107
  right middle lobe 92, 94, 95
  superior lobe 78, 92, 93, 94, 95
lymph nodes, 89
mamillary process 13
mandible 4, 7
  angle 7
  body 7
  coronoid process 7
  mental foramen 4, 7
  mental protuberance 7
  mental tubercle 7
  notch 7
  oblique line 7
  ramus 7
mandibular fossa 6
mastoid groove for digastric muscle 6
mastoid process 6
maxilla 4, 6
maxilla, orbital surface 4
medial collateral ligament 57
meniscus, lateral 57
meniscus, medial 57
mesenteric veins 83
metacarpus 24
metatarsus 25, 32, 58
midbrain 67
middle colic vein 103
middle cranial fossa 5
muscles and tendons
  abductor digiti minimi 50, 58
  abductor hallucis 58
  abductor pollicis brevis 50
  achilles tendon 52, 53, 55, 58
  adductor brevis 51, 56
  adductor longus 52, 53, 54, 56
  adductor magnus 51, 53, 54, 56
  adductor pollicis 50
  anconeus 45, 50
  biceps brachii 43, 44, 45, 46, 50
  biceps femoris 52, 53, 54, 55, 57
  brachialis 44, 48, 49, 50
  brachioradialis 43, 44, 47, 48, 49, 50
  buccinator 36
  deltoid 36, 38, 43, 44, 45, 46, 47
  depressor anguli oris 36
  diaphragm 111
  extensor carpi radialis longus 43, 44, 48-50
  extensor carpi ulnaris 43, 44, 50
  extensor digitorum 43, 44, 45, 50
  extensor digitorum longus 52, 55, 56, 58
  extensor hallucis longus tendon 58
  external abdominal oblique 38, 41, 51, 107
  external intercostal 42, 46
  external obturator 51
  flexor carpi radialis 50
  flexor carpi ulnaris 45
  flexor digitorum brevis 58
  flexor digitorum longus 53, 56
  flexor digitorum profundus 44, 45
  flexor digitorum superficialis 50
  flexor hallucis longus 56, 58
  flexor pollicis longus 48, 49
  gastrocnemius 52, 53 54, 55, 57
  gemellus inferior 51, 53, 56
  gemellus superior 51
  genioglossus and geniohyoid attachment 7
  gluteus maximus 38, 51, 52, 53, 54, 107
  gluteus medius 38, 51, 53, 54
  gluteus minimus 51, 53
  gracilis 51, 52, 53, 54, 55, 57
  iliacus 51, 52, 53, 54, 56
  iliocostalis 40
  iliocostalis muscle 107
  iliotibial tract 51, 52, 53, 54, 55
  inferior obliquus capitis 37
  infraspinatus 36, 38, 45, 46, 48, 49
  internal abdominal oblique 42, 51, 107, 111
  internal intercostal 46
  internal obturator 51, 53, 56
  interosseous 48, 49, 50, 58
  lateral pterygoid 7, 37

latissimus dorsi 38, 43, 44, 45, 46, 47, 111
levator scapulae 36
longissimus capitis 40
longissimus dorsi 107
longus capitis 37
longus colli 37
masseter 36
medial pterygoid 7, 37
multifidus 107
mylohyoid 7, 36
omohyoid 36
palmaris longus 50
pectineus 51 53 56
pectoralis major 36, 38, 41, 43, 44, 46, 47,107
pectoralis minor 48
peroneus brevis 56, 58
peroneus longus 52, 53, 54, 55, 56, 57, 58
piriformis 51, 56
plantaris 53, 54, 57
popliteus 57
popliteus tendon 57
pronator quadratus 48, 49
pronator teres 50
psoas major 51, 52, 53, 54, 56, 107, 111
quadratus femoris 51
quadratus lumborum 40, 107
quadratus plantae 58
quadriceps tendon 57
rectus abdominis 41, 42, 51, 107
rectus abdominis, tendinous intersection 41, 42
rectus capitis posterior major 37
rectus capitis posterior minor 37
rectus femoris 51, 52, 53, 54
rectus sheath 41, 42
rhomboideus major 49
rhomboideus minor 46, 48, 49
sartorius 51, 52, 53, 54, 55, 57
scaleni 36, 37, 40, 46
scalenus anterior 37
scalenus medius 37
scalenus posterior 37, 46
semimembranosus 52, 53, 54, 55, 57
semispinalis 40
semitendinosus 53, 54, 55, 57
serratus anterior 38, 41, 46, 47
soleus 52, 53, 55, 57, 58
spinalis 40
splenius capitis 36, 40
sternocleidomastoid 36, 38, 47
sternohyoid 36
sternothyroid 47
stylohyoid 36, 37
subclavius 46
subscapularis 46
supraspinatus 46, 48, 49
temporalis 36
tensor fasciae latae 51, 52, 54
teres major 36, 38, 46, 48, 49
teres minor 46
thoracolumbar fascia 38, 45
tibialis anterior 52, 54, 55, 57, 58
tibialis posterior 56, 58
transversus abdominis 42, 107, 111
trapezius 36, 38, 43, 44, 45, 46, 47
triceps brachii 36, 43, 45, 46, 47, 50
triceps brachii, long head 36
triceps brachii, scapular attachment 46
vastus intermedius 51
vastus lateralis 51, 52, 53, 54, 55
vastus medialis 52, 53, 54
mylohyoid line 7
nasal bone 4, 6
nasal spine, posterior 6
navicular 32, 58
nerves 58, 62, 63, 64, 67, 70, 71
common peroneal 71
femoral 71
medial sural cutaneous 71
median 70
musculocutaneous 70
optic 62, 63, 64, 67
radial 70
suprascapular 70
sciatic 71
tibial 58, 71
ulnar 70
nucleus accumbens 69
occipital condyle 6
occipital lobe 62, 67, 68
optic canal 5
chiasm 63, 67
optic recess 65
optic tract 67
orbital fissure, superior 4, 5
pancreas 98, 99, 103, 111, 117
body 103
head 103
neck 103
tail 103
pancreatic duct 111
paraterminal gyrus 64
parietal bone 4
parietal lobe 62, 68, 69
parotid glands 98, 99
patella 25, 28, 29, 57
patellar ligament 57
patellar retinaculum 52, 53, 54, 55, 57
pectoral girdle 3, 18
pelvic brim, asacral part 27
pelvic girdle 3, 9
penis 109, 110, 113, 114
bulb 114
corpus cavernosum 114
corpus spongiosum 114
crus 114
glans 109, 110, 113, 114
pericardium 78, 79
petrotympanic fissure 6
pharyngeal tubercle 6
pineal gland 117
pisiform bone 24
pituitary gland 117
pons 62, 63, 64, 69
popliteal artery and vein 57, 73, 74
portal vein 83, 86, 87, 107
posterior cranial fossa 5
posterior descending artery 75
posterior tibial artery and vein 58, 73, 74
primary bronchus 79, 92, 94, 96
prostate 109, 110, 113, 114
pterygoid fossa 6
pubic ramus, superior 26, 27
symphysis 26
tubercle 26, 27
pulmonary artery 75, 76, 77, 79, 84, 85
trunk 76, 78, 84, 85
vein 75, 79, 84, 85
putamen 67, 68, 69
radial tuberosity 22, 23
radius 18, 19, 22, 23
head 23
neck 23
rectum 98, 102
renal artery 83, 111
renal cortex 111
fat 111
medulla, pyramid 111
papilla 111
pelvis 111
renal vein 83, 86, 111
rib, ribs 2, 15, 46, 79, 111
cage 2
angle 17
articular facet for vertebral body 17
articular surface 16
costal groove 17
false 15
floating 15
head 17
neck 17
true 15
scaphoid bone 24
scapula 18, 19, 20, 36, 38, 45, 46
acromion 20, 46
body 46
coracoid process 20, 46
costal surface 20
inferior angle 20
infraspinatus fossa 21
spine 36, 38, 45, 46
suprascapular notch 20
supraspinatus fossa 20
septum pellucidum 69
sesamoid 24
skull 2
small intestine 98, 99, 111
spermatic cord 51, 113, 114
sphenoid 4, 6
greater wing 4
hamulus 6
lesser wing 4
spinal cord 60, 61, 62, 63, 64, 69, 111
spine 2, 9
spleen 83, 89, 90, 111
splenic vein 83, 90, 103, 111
sternum 15, 16, 79
angle 16
body 15, 16
jugular notch 16
manubrium 15, 16
xiphoid process 15,16
stomach 98, 99, 100
body 100
cardiac part 100
fundus 100
greater curvature 100
lesser curvature 100
pyloric antrum 100
pylorus 100
styloid process 6, 22, 23
stylomastoid process 6
subclavian artery (es) 75, 80, 81, 82
subclavian vein 82
submandibular glands 98, 99
superior mesenteric artery 111
superior pulmonary vein 75, 84, 85
supraorbital notch (foramen) 4
suprapatellar fat pad 57
talofibular ligament, anterior 58
talus 30, 32, 58
articular surface 30
tarsal bones 25, 30, 32, 58
temporal bone, styloid process 4
temporal lobe 62, 67, 68, 69
temporal pole 63
temporal process 4
testis (left only) 113, 114, 117
thalamus 67, 68, 69
thymus 89, 90
thyroid cartilage 92, 93, 94, 95, 96
thyroid gland 117
tibia 25, 30, 31, 57, 58
anterior border 31
anterior intercondylar area 31
anterior intercondylar tubercle 30
anterior surface 30
articular surfaces 30, 31
intercondylar eminence 30, 57
lateral articular surface 30
lateral condyle 30, 31, 57
lateral intercondylar tubercle 30
medial articular surface 30
medial condyle 30, 31, 57
medial intercondylar tubercle 30
medial malleolus 30, 31, 58
plateau 30
tuberosity 31
trachea 79, 92, 94, 95, 96
transverse tibiofibular ligament 58
trapezium 24
trapezoid 24
triquetral bone 24
ulna 18, 19, 22, 23
coronoid process 22, 23
olecranon 22, 23
radial notch 23
trochlear notch 22, 23
tuberosity 22, 23
notch 22
upper appendicular skeleton 2, 3
upper extremity 3, 34, 35
ureter 109, 110
urethra 109, 110, 114
membranous part 110, 114
spongy part 110
urinary bladder 51, 109, 110
vena cava 73-77, 79, 82, 83, 86, 87, 107, 111
inferior 73, 74, 75, 76, 77, 79, 82, 83, 86, 87, 107, 111
superior 73, 74, 75, 76, 77, 79, 82
ventricle 65, 67-69
fourth 65
lateral 65, 67, 68, 69
lateral, occipital horn 65
right lateral, frontal horn 65
third 65, 67, 69
vertebrae
cervical 8, 9, 10
C1 (atlas) 10
C2 (axis) 10
vertebral foramen 10
superior articular facet 10
superior articular process 12
coccyx 8, 9, 14, 26, 27
lumbar 8, 9, 13, 107, 111
inferior articular process 13
lamina 13
pedicle 13
transverse process 13, 111
spinous process 13
superior articular process 13
vertebral body 13, 111
sacrum 8, 9, 14, 26, 27
ala 14, 26, 27
anterior sacral foramen 14, 27
apex 27
articular surface 14, 27
transverse process 14
lateral part 27
lumbosacral articular surface 26-27
median sacral crest 14, 26, 27
promontory 14
sacral foramen, posterior 14, 26-27
sacral hiatus 14
sacral tuberosity 14, 26, 27
superior articular process 26, 27
superior articular surface 14
transverse ridge 14, 27
thoracic 8, 9, 12, 15, 17
facet for transverse process 17
inferior articular process 12
inferior vertebral notch 12
lamina 12
pedicle 12
spinous process 12, 15
superior costal facet 12
transverse costal facet 12
transverse process 12
vertebral body 12
vertebral artery 81
vertebral canal 13
vomer 4, 6
basilar part 6
zygomatic arch 6
zygomatic bone 4, 6
zygomatic process 6